THE FROG

THE FROG

AN INTRODUCTION TO

ANATOMY, HISTOLOGY, AND EMBRYOLOGY

BY THE LATE

A. MILNES MARSHALL., M.D., D.Sc., M.A., F.R.S.,

EDITED BY

F. W. GAMBLE, D.Sc., F.R.S.

ELEVENTH EDITION

GYAN

Published by

Gyan Publishing House
5, Ansari Road
Daryaganj, New Delhi-110002
Phone: 011-47034999, 9811692060
E-mail: books@gyanbooks.com

Distribution Network
gyanbooks.com
India, USA, Canada, UK, Australia, France

ISBN : 978-81-212-6053-4 (PB)
First Published, 1920

2nd Impression 2022

Printed at: Gyan Press, Delhi.

THE FROG
Edited by: F. W. GAMBLE

THE FROG:

AN INTRODUCTION TO

ANATOMY, HISTOLOGY, AND EMBRYOLOGY

BY THE LATE

A. MILNES MARSHALL, M.D., D.Sc., M.A., F.R.S.,

FORMERLY FELLOW OF ST. JOHN'S COLLEGE, CAMBRIDGE; PROFESSOR IN THE VICTORIA UNIVERSITY; BEYER PROFESSOR OF ZOOLOGY IN OWENS COLLEGE, MANCHESTER.

EDITED BY

F. W. GAMBLE, D.Sc., F.R.S.

MASON PROFESSOR OF ZOOLOGY AND COMPARATIVE ANATOMY UNIVERSITY OF BIRMINGHAM

ELEVENTH EDITION

1920

PREFACE TO THE ELEVENTH EDITION.

THIS little work is intended to supply the student with a practical guide to the study of elementary anatomy, histology, and embryology. For this purpose the frog is the animal chosen, as being easy to obtain, convenient to dissect, and a fairly typical example of the great group of vertebrate animals. Where, from its small size, or for other reasons, the frog has proved unsuitable, other animals have been substituted for it. For convenience of reference, and in order definitely to stamp the practical character of the work, directions for dissection, etc., have been printed in italics.

In preparing this edition a few minor alterations have been made in the text, and a section on cell-division and the development of germ-cells has been added.

F. W. GAMBLE.

BIRMINGHAM, *January* 1912.

CONTENTS.

CHAPTER VI.

The Eye and Ear.

CHAPTER VII.

The Reproductive Organs and the Cloaca.

CHAPTER VIII.

Development of the Frog.

CHAPTER IX.

Elementary Histology.

CHAPTER X.

Cell Division : Development of Germ-Cells.

LIST OF ILLUSTRATIONS.

INTRODUCTION.

I.—LIST OF APPARATUS REQUIRED.

THE following apparatus is recommended to the student of Elementary Biology:

1. Two or three scalpels or dissecting knives of different sizes.

2. Two pairs of forceps, one large and one small. Both pairs should be straight, and should have the tips roughened in order to secure a firmer hold.

3. Two pairs at least of scissors; one pair large and strong, for cutting bone and other hard tissues; the other pair small, for fine dissections. A second small pair may have the blades bent at an angle (elbow scissors). In selecting scissors be careful to see that they cut quite up to the points of the blades.

4. A pair of stout needles, firmly mounted in handles.

5. A pair of the finest sewing needles, mounted in handles: only about a quarter of an inch of the needle should project. They are used for teasing histological preparations.

6. A seeker, *i.e.*, a blunt needle mounted in a handle, and bent at an angle half an inch from the end.

7. A metal blow-pipe: and a glass cannula with india-rubber cap.

8. A pocket lens, containing two or three lenses mounted in a handle, and giving when combined a magnifying power of at least six diameters. A strong "watchmaker's lens" is most useful to those who can hold it at their eye, as leaving both hands free to dissect.

9. Slides and coverslips, for mounting microscopical specimens. The coverslips should be the thinnest sold (No. 1). Square covers are easier to handle than circular.

10. A blank note-book, for drawing in; an HB pencil, and a piece of india-rubber.

11. A cheap pair of compasses, for measuring the dissections.

II.—ON DISSECTION.

The object of dissection is to separate the several parts and organs from one another, so as to define their boundaries and display clearly their mutual relations. Dissection consists mainly in removing the "connective tissue" which binds the several parts together.

The following rules should be carefully attended to:

1. Pin down the animal firmly to the dissecting board. Never attempt to dissect a specimen that is not so fixed.

2. In pinning out a dissection stick the pins in, not vertically, but obliquely, so that their heads do not get in the way or obscure the dissection.

3. Never cut away anything until you are quite certain what it is you are removing.

4. Put the part you are dissecting slightly on the stretch; *e.g.*, when dissecting the bloodvessels or nerves of the throat, distend it by passing a small roll of paper or lumps of cotton-wool down the œsophagus; or when dissecting the muscles of the leg, pin out the leg in such a position as to stretch the muscles you are cleaning.

5. In cleaning bloodvessels or nerves always dissect along them, and not across them; and avoid laying hold of them with the forceps. Similarly when cleaning muscles, dissect along their fibres and not across them.

6. Fine dissections should be done under water, which supports the parts and greatly facilitates the operation. A stream of water allowed to play gently on the dissection from time to time is often a valuable aid.

7. The dissection of muscles, and still more of nerves, is greatly aided by placing the specimens in spirit for a day before dissecting.

8. Keep your instruments clean and sharp. Be careful not to blunt your fine scissors or scalpel by using them for cutting hard parts.

9. If you get in a muddle, stop and wash the dissection thoroughly under the tap before proceeding further.

III.—ON DRAWING.

It is absolutely essential to draw your dissections, and this must on no account be omitted. Keep a separate book for your drawings, and draw every dissection you make. Do not be discouraged if you find it difficult at first: you will never regret time spent on it.

The following rules will be useful to those who have not learnt to draw systematically:

1. Make your drawing to scale, *i.e.*, either the exact size of the natural object, or half or double or treble that size, as the case may be, remembering always that a drawing can hardly be made too large.

2. In commencing a drawing, first determine by careful measurement the positions of the principal points, and sketch in lightly the whole outline before finishing any one part.

3. If the object you are drawing is bilaterally symmetrical, draw a faint line down the middle of your paper, and sketch in the left-hand half first; by measuring from your median line it will be very easy to make the two halves symmetrical.

4. Name on your drawing the several parts shown, and mark also the scale adopted. If your drawing be of the natural size mark it thus— $\times 1$; if it be double the size of the object mark it $\times 2$; if half the size, $\times \frac{1}{2}$, and so on.

5. Draw on one side of the page only: and write an explanation of your drawing on the opposite page.

6. Always make your drawing in pencil first, since much clearer outlines can be obtained with pencil than with chalk, but for complicated drawings coloured pencils are very useful, and water-colour paints still better. Keep certain colours for particular organs or tissues; *e.g.*, when drawing the skeleton colour the cartilage blue, the cartilage bones yellow, and the membrane bones either red or white; when drawing the blood-vessels colour the arteries red and the veins blue.

7. Draw only what you see.

IV.—THE USE OF THE MICROSCOPE

The microscope consists essentially of a *stand* and a *body*, the latter of which bears at its ends the *lenses* by which the magnifying power is obtained.

The stand is an upright pillar, the lower end of which is attached to a heavy foot to ensure steadiness. A little way above the foot the stand supports a horizontal plate—*the stage*—on which the object to be examined is placed. The stage is perforated in the middle by a hole, the size of which can be varied by means of diaphragms. Through this hole light is directed on the object to be examined by means of a mirror attached to the stand below the stage. Above the stage the stand supports a vertical tube, in which the body of the microscope slides up and down.

The body is a tube, in the upper end of which is placed a combination of lenses, known as the *eyepiece*, while to the lower end is screwed another combination of lenses—the *objective.*

A microscope is usually provided with a couple of eyepieces and a couple of objectives of different magnifying power. An objective magnifying only a small number of times is called a *low power;* one magnifying many times (200 diameters or more), a *high power.* Similarly eyepieces are spoken of as high or low according to their magnifying power.

In order that an object may be seen clearly the objective must be at a certain definite distance from the object, this distance (the "focal length") varying with different objectives, and to a slight extent with different observers. The higher the power employed the closer must the objective be brought to the object. As the position of the object on the stage of the microscope is fixed, this distance is regulated by moving the body of the microscope up and down in the tube in which it slides.

This process of *focussing* is effected in two ways:

(1) By simply sliding the body up and down by hand in the stand, or by screwing it up and down with a rack and pinion, according to the type of microscope employed. This is known as the *coarse adjustment.* The sliding of the body should be performed with a slightly screwing motion, and can only be used when low powers are being employed.

(2) With high powers the objective has to be brought so close to the object that a more delicate method of adjustment is necessary. This *fine adjustment* is effected by a screw with a milled head placed at the top of the vertical pillar forming the stand. By turning the head from right to left, in the direction of the hands of a watch, the body of the microscope is lowered and the objective brought nearer to the object: by turning in the reverse direction the objective is raised.

In using the microscope attend to the following rules:

1. Always examine an object first with the low power. Having adjusted the eyepiece and objective, direct the light up the tube

of the microscope by means of the mirror, and then place the object on the stage. Twist down the body until the objective is about a quarter of an inch from the cover-glass; look down the microscope, and gradually twist the body up until the object becomes visible. Focus accurately by means of the fine adjustment screw.

2. When using a high power begin with the objective close to the cover-glass, and then focus with the fine adjustment. It will facilitate the process if, while focussing with the right hand, you move the object about slightly with the left hand.

3. Take extreme care never to let the objective touch the cover-glass; and never to touch or allow any dirt to get on the face of the objective. The face of an objective cannot be cleaned without doing harm to it.

4. Should by any chance a drop of glycerine get on the face of the objective, wash it carefully with water from a wash-bottle, and wipe it gently with a silk handkerchief or piece of chamois leather. Should Canada balsam be allowed to get on the objective, do not attempt to clean it yourself, but hand it at once to the assistant.

5. See that the body of the microscope slides smoothly in its tube. If it does not, remove it, and clean it by rubbing with a few drops of olive oil: wipe off the oil before replacing the body in the tube.

6. Keep both eyes open when looking through the microscope: a very little practice will enable you to do this, and it will save you much fatigue. Also get into the habit of using either eye.

7. With a high power, use a small diaphragm: the amount of light will be somewhat diminished, but the clearness and definition of the object much increased.

8. When examining an object, keep one hand on the fine adjustment, and keep screwing it up and down slightly the whole time: in this way parts of the object at different depths are brought into focus successively, and a clearer idea of the object is obtained.

9. If the object appears dim or dirty, find out where the fault lies in this way:

While looking down the microscope, turn round the eyepiece with your right hand. If the dirt turns round too, remove and clean the eyepiece. If the fault is not in the eyepiece, move the slide about gently; if the dirt moves with the slide, remove the slide and clean it. If the dirt does not move with either the eyepiece or the slide the fault is almost certainly in the objective, which should be removed and examined; if dirty, it must be cleaned very carefully with a piece of silk or chamois leather.

V.—THE PREPARATION OF MICROSCOPICAL OBJECTS.

In mounting microscopical objects be careful that your slides and coverslips are thoroughly clean. Slides should be labelled as soon as they are prepared, and should be kept in a box or cabinet in which they lie flat.

A. Methods of Mounting.

There are various media in which objects may be mounted. The method of procedure is much the same with all. Put a small drop of the fluid in the middle of the slide, place the object in the middle of the drop, and arrange it with needles in the position desired. Then place the cover-glass carefully on the top, letting it rest by one edge on the slide and supporting the opposite edge by a needle: withdraw the needle gradually so as to let the cover-glass down slowly, and drive out any air-bubbles there may be in the fluid. If any air-bubbles still remain, leave them alone, as they will probably work out by themselves. Be careful not to use too large a drop of your mounting medium, and above all things be careful not to let any of it get on the top of the cover-glass; should this happen, the cover-glass must be removed at once and the specimen mounted afresh with a clean one.

The most important mounting media are the following:

1. **Normal Salt Solution**: a 0·75 per cent. solution of common salt in water. This is very useful in the examination of fresh specimens of animal tissues, as, unlike water, it has practically no action on them. It cannot be used, however, for making permanent preparations.

2. **Glycerine** can be used either pure or diluted with its own bulk of water. If the preparations are intended to be permanent, a narrow ring of cement must be painted round the edge of the cover-glass to fix it to the slide. For permanent

preparations it is better to use glycerine jelly; a drop of this should be melted on the slide, and the object transferred to it from glycerine: ring with cement as before.

3. **Canada Balsam** is the most generally useful medium for permanent preparations, as requiring no cement. Specimens that are to be mounted in balsam must first be deprived of all water they may contain by placing for an hour or so in absolute alcohol, and should then, before mounting, be soaked for a few minutes in oil of cloves or turpentine in order to clear them, *i.e.*, render them permeable by the balsam. Canada balsam, if too thick, may be diluted with chloroform or benzole.

B. Teasing.

The object of teasing is to separate the several parts of a tissue or organ from one another in order to show their minute structure.

The fragment to be teased should be placed on a slide in a drop of the medium in which it is to be mounted, and then torn up into shreds by means of a couple of needles held one in each hand. The process is often greatly facilitated by placing the slide on a piece of black paper, which renders the particles easier to see. When torn up as finely as possible, a cover-glass is placed on as before. The two rules to be borne in mind in teasing are the following:

1. Take a *very small* fragment to commence with; hold it with one needle, and tear it with the other.
2. Tease it as finely as you can. Your object is to separate the component parts from one another.

C. Maceration.

The process of teasing is in many cases facilitated by previously macerating the specimen, *i.e.*, soaking it in some fluid, which, while preserving the individual cells, tends to loosen them from one another. The most important macerating fluids are as follows:

1. **Ranvier's Alcohol**: a mixture of one part of strong spirit with two parts of water. The specimen should be put fresh into the mixture and allowed to remain twenty-four hours or more.

2. Müller's Fluid: a solution of bichromate of potash with a little sodic sulphate in water.

D. Staining.

Various reagents are employed for the purpose of staining preparations; some of these merely dye the whole preparation more or less uniformly, but the most useful ones are those which stain certain parts of the cells only, or at any rate stain these much more strongly than the other parts (selective stains). The most important are the following:

1. Hæmatoxylin. There are various preparations of hæmatoxylin, or logwood, used in microscopical work: the best is that proposed by Delafield. It is prepared thus: dissolve 4 grammes of crystallised hæmatoxylin in 25 cubic centimetres of strong alcohol; add this solution to 400 c.c. of a saturated solution of ammonia alum, and expose to the light in an unstoppered bottle for 3 to 4 days. Filter, add 100 c.c. glycerine and 100 c.c. of 90 per cent. alcohol.

The specimens, which must be perfectly free from all trace of acid, should be cut into small pieces, and passed through weak spirit to water. They should then be left in the hæmatoxylin in a covered vessel or stoppered bottle for from one to twelve hours, according to the size of the specimen and the depth of staining desired, and then brought up through water and weak spirit, and left in strong spirit for some hours before mounting. Hæmatoxylin stains the nuclei of cells much more strongly than the other parts.

2. Borax-Carmine. This, which is perhaps the most generally useful of all the stains in ordinary use, is prepared as follows. Dissolve 2 parts of carmine and 4 parts of borax in 100 parts of water: add an equal volume of 70 per cent. alcohol; let the mixture stand for a couple of days, and then filter.

Specimens may be left in borax-carmine for from one to twenty-four hours, or even for two or three days: on removal they should be placed in **acid-alcohol**—*i.e.*, 70 per cent. alcohol to which a few drops of hydrochloric acid have been added—until they become a bright scarlet colour, when they should be transferred to 70, and then to 90 per cent. alcohol, in which latter they may be left till required. The time of immersion in acid-alcohol will vary, according to the nature and size of the specimen, from a quarter of an hour up to a day or more.

3. **Picro-Carmine** is a very useful, and to a certain extent a differential stain, and it colours the several tissues different tints. It may be prepared thus. Dissolve 1 gramme of carmine in 4 c.c. of liquid ammonia and 200 c.c. of distilled water. Add 5 grammes of picric acid; shake the mixture well for some minutes, and then decant from the excess of acid. Leave the decanted liquid for some days, stirring it occasionally: then evaporate it to dryness, and to every 2 grammes of the dried residue add 100 c.c. of distilled water.

Picro-carmine answers best with specimens preserved in 70 per cent. alcohol. They should be brought through weak spirit to water, and left in the stain for a day; then brought up to 70, and afterwards to 90 per cent. alcohol. Some specimens give better results when washed freely with water on removal from the picro-carmine, and then placed in 1 per cent. acetic acid for an hour before transferring to alcohol.

4. **Magenta** stains very rapidly but diffusely: the colour also is not permanent.

5. **Silver Nitrate.** A ½ per cent. solution in water stains the intercellular substance, which binds together the several cells of a tissue, much more strongly than the cells themselves, and is therefore chiefly used when we wish to render prominent the outlines of the individual cells. The specimens should be placed fresh in the silver solution for from two minutes to a quarter of an hour, then washed thoroughly with distilled water, and exposed to the light until stained sufficiently deeply, when they may be mounted in glycerine. Such preparations are rarely permanent, as the reduction of the silver, to which the staining is due, continues until the specimens ultimately become too dark to be of any use.

6. **Osmic Acid.** A 1 per cent. solution of osmic acid in water forms an extremely useful staining reagent. It is especially useful for the detection of fat, which is stained by it a dark brown or black colour. Specimens, which must be quite fresh, should only be left in it a few minutes, and may then be mounted in glycerine, or else washed, dehydrated, and mounted in balsam.

7. **Acetic Acid.** Although not strictly a staining agent, inasmuch as it does not colour the specimens, acetic acid may

conveniently be mentioned here, as it is used for the same purpose as the true stains, *i.e.*, for the sake of rendering certain parts of the cells especially distinct. Acetic acid, of which a 1 per cent. solution is employed, causes the protoplasm of cells to swell up and become transparent, and brings the nuclei into special prominence. It is used with fresh specimens.

VI.—ON PRESERVING AND HARDENING.

The reagents in common use for killing and preserving small animals are valuable also from their power of "fixing" the tissues, *i.e.*, of coagulating the protoplasm of the cells. The objects to be attained are to effect this coagulation quickly, before the tissues can undergo any alteration, and thoroughly, *i.e.*, throughout the whole thickness of the object to be hardened. They are as follows:

1. **Alcohol.** Specimens may be placed at once in 70 per cent. alcohol; and thence transferred after a couple of days to 90 per cent., in which they may be left till required.

2. **Osmic Acid.** For this purpose a 1 per cent. solution in water is used: it acts almost instantaneously, and so allows no change to occur in the tissues; it has also the merit of staining the tissues as well as hardening them. It can, however, only be employed when the specimens are very small, as it hardens the surface layers so rapidly that it is unable to penetrate beyond a very slight depth. A few minutes' immersion is usually sufficient. The vapour must not be inhaled.

3. **Corrosive Sublimate.** This is by far the best general reagent for killing and fixing small animals. A saturated solution in water is employed, in which the object is placed for half an hour or more.

4. **Chromic Acid.** A 0·25 to 0·5 per cent. solution of chromic acid in water is a useful hardening reagent; it acts much more slowly than osmic acid, but penetrates to greater depths. Specimens should usually be left in the solution for one or more days, and then thoroughly washed for some days in increasing strengths of alcohol. These processes should be carried out as far as possible in a dark drawer or cupboard.

5. **A Mixture** of chromic acid with a few drops of osmic acid

is often very useful, as it combines the advantages of both reagents.

6. **Picric Acid** is a very valuable hardening reagent, of which the best preparation is Kleinenberg's. Specimens should be left in it from 12 to 24 hours. It is prepared thus: with 100 c.c. of water make a cold saturated solution of picric acid: add 2 c.c. of concentrated nitric acid: filter, and add to the filtrate three times its volume of water.

VII.—ON WASHING.

Specimens that have been hardened in any of the preceding reagents (except alcohol) should be placed in water on removal from the reagent, then in 30 per cent. alcohol, then in 50 per cent. alcohol. They should then be washed in alternate strengths of 50 per cent. and 70 per cent. alcohol in order to remove all traces of the hardening reagent, and should be stored in 70 per cent. alcohol. The object of using graduated strengths of alcohol is to avoid distortion of the tissues by too sudden changes of specific gravity. These remarks apply also to treatment before and after staining.

VIII.—ON SECTION CUTTING.

Many tissues and organs can only be studied satisfactorily by cutting them into thin sections, and this method of investigation is of such importance as to require special notice. There are three chief stages: Hardening, Imbedding, and Cutting, which will be noticed in succession.

A. Hardening.

Before the object can be cut into sections it is necessary to harden it; this may be effected by freezing, but the more usual plan is by means of reagents, which have been discussed under previous headings.

B. Staining.

The hardened specimens, if not too large, may now be stained with either hæmatoxylin, borax-carmine, or picro-carmine; they should then be brought to 70 or 90 per cent. alcohol. If the specimen is too large to stain whole, the sections must be stained after they are cut (*see* p. 13).

C. Imbedding.

The preparation of sections is greatly facilitated by imbedding the specimen in some waxy substance. For this purpose various materials have been employed, but by far the most useful is paraffin-wax, which is used in the following manner:

The stained specimen is placed in absolute alcohol for an hour or two in order to completely dehydrate it. It is then transferred to turpentine or benzole, in which it is left for half an hour or more until completely saturated. From the turpentine or benzole it is transferred to melted paraffin, which is kept by means of a water-bath at a temperature just above its melting-point (about 56° C.). In this it is left for several hours, or even for a whole day, in order that it may be thoroughly permeated. It is then placed in a small box of paper, or other material, filled with melted paraffin. By means of hot needles it can readily be arranged in any desired position; and the paraffin should then be cooled quickly.

D. Section Cutting.

When thoroughly set the block is removed from the box, and the paraffin pared away with a knife until the specimen just comes into view.

The block is then placed in a microtome, and cut into thin sections. These may be transferred one by one to a slide, but a great saving of time is effected by the method of cutting continuous ribbons, devised by Mr. Caldwell.

This depends on the fact that if the paraffin is of proper consistency the successive sections, as they are cut, will stick together at their edges, so as to form a ribbon. To ensure this the edge of the razor should be placed at right angles to the direction of stroke, and the edges of the block of paraffin cut parallel to one another, and to the edge of the razor. If for any reason it is desirable to imbed the specimens in a paraffin too hard to form ribbons, the block should, before cutting, be coated with a layer of soft paraffin, by dipping it for a few moments in a dish of melted soft paraffin. This outer coating should be left on the sides of the block parallel to the edge of the razor, but cut away from the sides at right angles to it.

The razor should be used dry: and the sections, when cut, placed on slides painted, just before they are used, with a thin layer of a mixture of collodion and oil of cloves in equal parts. The slide is then heated by a water-bath to a temperature not exceeding 55° to 60° C., so as to melt the paraffin and evaporate the oil of cloves. The melted paraffin should then be washed off by turpentine, when the sections will remain fixed to the slide by the collodion, and may be mounted in balsam in the usual manner.

Instead of the mixture of collodion and oil of cloves, a solution of shellac in absolute alcohol may be used: this should be spread over the slide in a thin layer by means of a glass rod, and allowed to dry. Immediately before being used the slide should be brushed over with oil of cloves.

If it is desired to stain the sections on the slide, the thinnest possible layer of glycerine and albumen (pure glycerine and fresh white of egg in equal parts, filtered) should be smeared over the slide, and the sections laid on it in order. The paraffin, when melted, should be washed off with turpentine or benzole, and the sections dehydrated with absolute alcohol; the desired stain can then be used in the usual manner. Dehydrate with absolute alcohol, followed by oil of cloves, before mounting in Canada balsam.

VIII.—TABLE OF THE CHIEF PROCESSES ALREADY DESCRIBED.

A. Killing and Fixing.

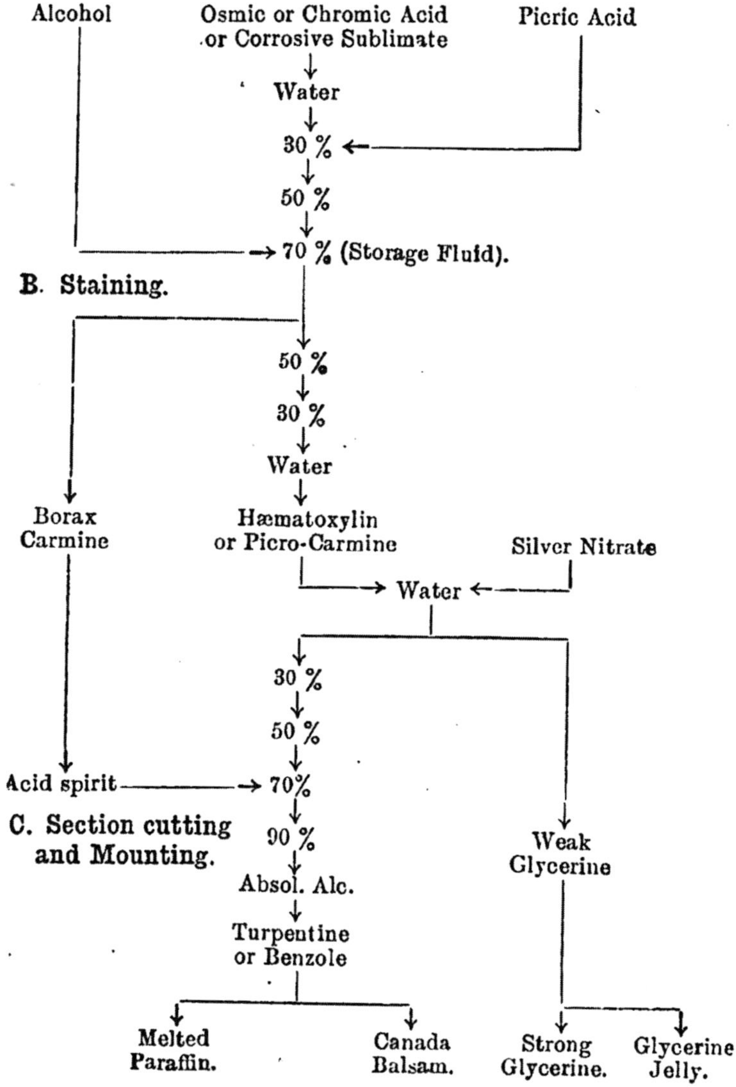

CHAPTER I.

GENERAL ANATOMY OF THE FROG.

FIG. 1.—The Common Frog (*Rana temporaria*) (from Ecker).

A. External Characters.

Lay the frog on a board before you; note, and make drawings showing the following points:

1. The division into head, trunk, and limbs; and the absence of neck and tail.

2. The two great surfaces.
   a. **The dorsal surface,** or back, is directed upwards when the frog is in the natural position.
   b. **The ventral surface,** or belly, is directed downwards towards the ground.

3. **The skin** is moist and smooth; and devoid of hairs, scales, and claws. The colour of the skin is variable in different specimens and at different times: it is mottled on the dorsal surface, paler on the ventral.

4. **The head** is flat and triangular, with a blunt apex directed forwards.

At the sides of the head are the eyes, which are large and prominent. Each eye has two eyelids, of which the upper is thick, pigmented, and almost immovable, while the lower is semi-transparent and freely movable.

Behind the eye on either side is an obliquely placed elongated patch of a dark colour, in the middle of which is a circular area—the **tympanic membrane**—supported by a firm marginal ring.

5. **The limbs.** There are two pairs of limbs, fore and hind; each limb being composed of three segments.

a. **The Fore limb** presents the following divisions:

i. **Arm.**

ii. **Forearm.**

iii. **Hand,** with four digits, corresponding to the four fingers of man; the thumb being very small and inconspicuous. In the male frog there is a pigmented thickening along the inner edge of the first digit, specially developed at the breeding season.

b. **The Hind limb** is much longer than the fore limb, and divided into the following parts:

i. **Thigh.**

ii. **Leg.**

iii. **Foot,** with five toes webbed together. The shortest toe corresponds to the big toe of man, and the longest to his fourth toe.

6. **External apertures**: or openings on the surface of the body.

a. **Median apertures.**

i. **The Mouth** is a wide horizontal slit.

ii. **The Cloacal aperture** is a small hole at the posterior end of the body, between the legs: it lies slightly on the dorsal surface, just behind the bony projection formed by the posterior end of the urostyle.

b. Paired apertures.

i. The **Nostrils** or **anterior nares** are two small openings on the dorsal surface of the head, close to its anterior end.

B. The Buccal Cavity.

Open the mouth to its full extent: note the wide buccal or mouth cavity, of which the hinder part or pharynx is continued back into the œsophagus. Note also the following structures:

1. On the Roof of the Mouth.

a. The Teeth.

i. The **maxillary teeth** are a row of fine teeth, attached round the edge of the upper jaw to the maxillæ and pre-maxillæ.

ii. The **vomerine teeth** are two small patches of sharp teeth in the fore part of the roof of the mouth and near the middle line, attached to the vomers.

b. The **posterior nares** are two small holes lying to the outer sides of and slightly in front of the two patches of vomerine teeth.

Pass bristles through the nostrils, and see that they come out through the posterior nares into the buccal cavity.

c. The **Eustachian tubes or recesses** are a pair of much larger holes, at the sides of the posterior part of the buccal cavity. Each hole opens into a slightly dilated chamber—the tympanic cavity—which is closed externally by the tympanic membrane already seen on the surface of the head.

Perforate the tympanic membrane on one side with a needle, and pass a bristle or seeker through the hole and down the Eustachian tube into the mouth.

d. Two rounded prominences at the sides of the roof of the mouth are caused by the eyeballs.

Press down one of the eyes with your finger, and note that it can be made to project very considerably into the buccal cavity.

2. On the Floor of the Mouth.

a. The **lower jaw**, which is devoid of teeth, forms a bony margin to the floor of the mouth: the rest of the floor is soft and fleshy, but is slightly stiffened by a cartilaginous plate—the body of the hyoid.

b. **The tongue,** which is thin and fleshy, is attached to the front part of the floor of the mouth, and has its free bilobed end turned backwards towards the throat.

Turn the tongue forwards with the forceps.

c. **The glottis,** or aperture of the larynx, is a longitudinal slit in the floor of the posterior part of the mouth, and is stiffened laterally by the arytenoid cartilages.

Pass bristles through the glottis into the lungs. If any difficulty is experienced in finding the glottis snip through the angles of the mouth with scissors, so as to allow the mouth to be opened more widely.

C. The Abdominal Viscera.

Lay the frog on its back under water, and fasten it down to the dissecting board by pins through the limbs. Cut through the skin, along the middle line, the whole length of the ventral surface. Separate the skin from the underlying parts, noticing its very loose attachment to these parts, and the large space—a lymph cavity—beneath it. Turn the flaps of skin outwards, and pin them back. Notice:

a. The muscles of the body-wall.

b. The **pectoral or shoulder girdle**: a bony arch running across the body, opposite the fore limbs.

Pinch up with forceps the muscular body-wall, and cut through it into the body-cavity or cœlom with scissors a little to each side of the median line, being careful not to injure the anterior abdominal vein which runs along the inner surface of the body-wall in that line.

Continue the cut of one side backwards to the hinder end of the body, and forwards to the jaw, cutting through the pectoral girdle with strong scissors, and taking care not to injure the parts beneath.

Tie ligature-thread twice round the anterior abdominal vein, cut between the two ligatures, and turn back the two halves of the vein in the body-wall.

Inflate the lungs with a blow-pipe through the glottis, and inflate the bladder through the cloacal aperture.

Note and draw the general arrangement of the viscera, showing the following structures:

1. **The heart,** enclosed in the **pericardium,** is situated in the middle ventral line, and in the natural condition of the parts is covered by the pectoral girdle and the sternum.

2. **The liver** is a large reddish-brown bilobed organ, behind and at the sides of the heart.

3. **The lungs** are two thin-walled elastic sacs at the sides of the heart: they lie dorsal to the liver, and are often hidden by it.

Note the bristles already passed into the lungs through the glottis.

4. **The small intestine** is a light-coloured convoluted tube; in the middle line behind is the much wider **large intestine.**

5. **The bladder** is a thin-walled bilobed sac at the posterior end of the body cavity.

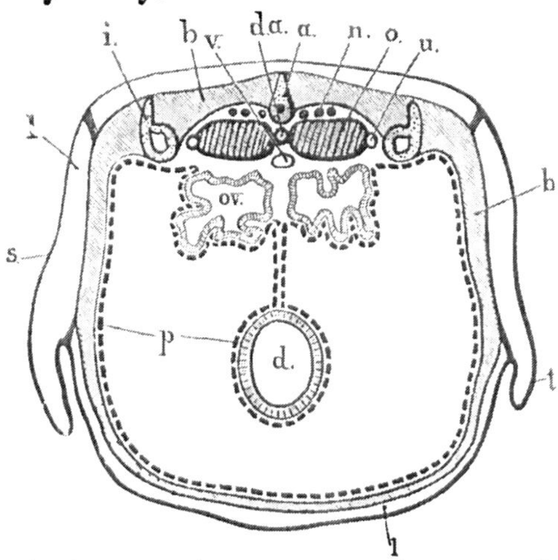

FIG. 2.—A diagrammatic transverse section across the posterior part of the body of a female frog.

a, urostyle; *b*, muscles of body wall; *d*, large intestine; *d.a*, dorsal aorta; *i*, ilium; *l*, lymph space between the skin and the muscular body-wall (cisterna lymphatica magna); *n*, spinal nerves; *o*, kidney; *ov*, oviduct; *p*, peritoneum; *s*, skin; *t*, fold of skin at groin; *u*, ureter; *v*, posterior vena cava.

6. *In the female frog note, in addition to the above parts,*

   a. **The ovaries:** two large bodies of irregular shape, each consisting of a mass of spherical black and white eggs, like small shot.

b. **The oviducts**: two long, very much convoluted tubes with thick white walls, lying at the sides of the body cavity.

7. *In the male frog note,*

a. **The testes**: a pair of ovoid bodies of a pale yellow colour, attached to the dorsal wall of the body cavity.

Notice the thin pigmented membrane—the **peritoneum**—which lines the body cavity or cœlom, covered by cœlomic epithelium. Trace this to the mid-dorsal line, where it is reflected downwards as a double layer—the **mesentery**—which embraces at its edge the alimentary canal, and binds its several coils together. (See Fig. 2.)

Notice also that all the abdominal viscera are really outside the peritoneum, which forms a closed sac into which the viscera are as it were pushed from without.

E. The Digestive Organs.

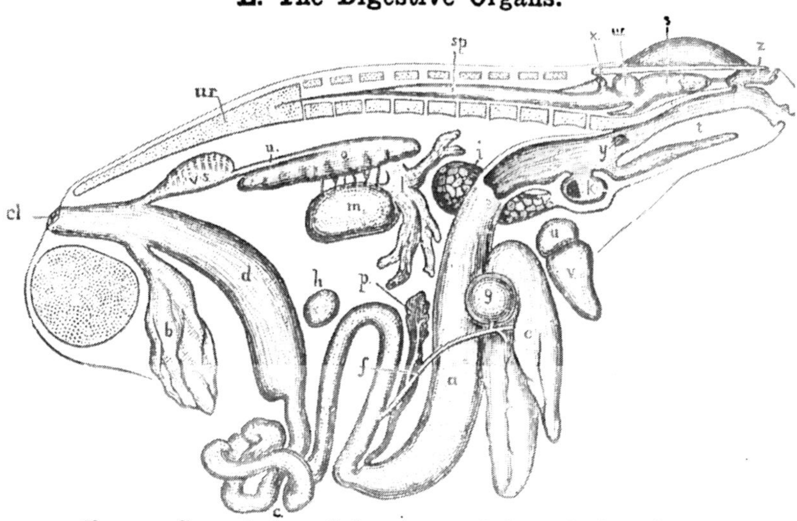

FIG. 3.—General view of the viscera of the male frog, from the right side.

a, stomach; *b*, bladder; *c*, small intestine; *cl*, cloacal aperture; *d*, large intestine; *e*, liver; *f*, bile duct; *g*, gall bladder; *h*, spleen; *i*, lung; *k*, larynx; *l*, fat body; *m*, testis; *n*, ureter; *o*, kidney; *p*, pancreas; *r*, pelvic symphysis; *s*. cerebral hemisphere; *sp*, spinal cord; *t*, tongue; *u*, auricle; *ur*, urostyle; *v*, ventricle; *v.s*, vesicula seminalis; *w*, optic lobe; *x*, cerebellum; *y*, Eustachian recess; *z*, nasal sac.

Turn the liver forwards, and note the stomach lying beneath its left lobe. Pass the handle of a seeker through the mouth and down the œsophagus into the stomach.

[*If the specimen be a female, remove the ovaries and oviducts completely, taking care not to damage the alimentary canal*]

1. **The Alimentary Canal.**

   a. **The œsophagus** is a short wide tube leading from the buccal cavity to the stomach.

   b. **The stomach** is a wide tubular sac about an inch and a half in length: it is narrower behind, and separated from the duodenum by a distinct **pyloric constriction.**

Cut open the stomach longitudinally along its left side, and wash out its contents: note the handle of the seeker already inserted through the mouth; also the longitudinal folds of the mucous membrane lining the stomach, which increase the extent of its surface.

   c. **The duodenum** is the first part of the intestine, rather more than an inch in length: beyond the pylorus it is bent back so as to lie parallel to the stomach. At its further end it is continuous with the small intestine.

   d. **The small intestine** is a slender convoluted tube about four and a half inches long, opening at its distal end by a small orifice into the large intestine.

   e. **The large intestine** is a short straight tube about an inch and a quarter long: it is very much wider than the small intestine, and opens behind to the exterior at the cloacal aperture.

   f. **The cloaca** in the frog is continuous with the large intestine, into it open the renal and genital ducts as well as the bladder: it will be described more fully when considering the urinary and reproductive organs. (See Chap. VIII.)

2. **The Liver.**

The liver is a large reddish-brown organ, divided into right and left lobes, connected together by a narrow bridge of liver-tissue. Of the two lobes the left one is much the larger, and is again subdivided into two.

a. **The gall-bladder** is a small spherical greenish sac lying between the right and left lobes of the liver.

b. **The bile duct** is a slender tube leading from the liver and gall-bladder to the duodenum, into which it opens about half an inch beyond the pylorus, and on the inner or concave side of the loop formed by the duodenum and stomach. The distal half of the bile duct traverses the pancreas: it has rather thick white walls and is easy to see; the upper half is more slender and more difficult to trace.

To trace the bile duct turn the liver forwards so that the point of attachment of the gall-bladder is clearly seen; and slightly stretch the duodenum by a pin passed through the pylorus. Determine the position of the two ends of the bile duct from the description given above, and dissect with a scalpel along the line thus indicated.

To see the opening of the bile duct, slit up the first three quarters of an inch of the duodenum along its convex border and wash out its contents: squeeze the gall-bladder so as to drive the bile along the duct into the duodenum: note the point at which it enters, and insert a bristle through the opening into the duct. Notice also the strong wavy transverse folds of the mucous membrane of the duodenum.

3. The Pancreas.

The pancreas is a whitish irregularly lobed mass lying in the loop between the stomach and duodenum, and best seen by turning the whole loop forwards. The pancreatic ducts are numerous and open into the bile duct, which passes through the pancreas to reach the duodenum.

Cut through the mesentery along its attachment to the intestine: uncoil the intestine, leaving it attached at both ends, and spread it out on your dissecting board: measure the lengths of the several portions and draw them to scale.

F. Other Abdominal Viscera.

1. The Kidneys are two flat elongated oval bodies of a red colour attached to the dorsal body-wall, close to the middle line, one on each side of the backbone or vertebral column. They lie in the large lymph space behind the peritoneum, and, like the other viscera, are outside the abdominal cœlomic cavity. (See Fig. 2, p. 19.)

a. **The ureters**, or ducts of the kidneys, are a pair of white tubes arising from the outer edges of the kidneys at about a quarter of their length from their hinder ends, and running back to open into the dorsal wall of the cloaca, opposite the opening of the bladder.

In the male frog a pouch-like dilatation, the **vesicula seminalis**, is present on the outer side of each ureter, close to its opening into the cloaca.

b. **The adrenal bodies** are small yellowish-red patches on the ventral surface of the kidneys.

c. **The corpora adiposa**, or **fat bodies**, are two bright yellow tufts of flattened processes attached to the dorsal wall of the body cavity; they vary much in size, and usually come to the surface just behind the liver.

2. **The Spleen** is a small round dark-red body lying in the mesentery, opposite the commencement of the large intestine.

G. Thyroid and Thymus.

1. The **thyroid** is a small paired gland attached to the base of the external jugular vein, a little way in front of the heart.

2. The **thymus** is a small ovoid paired gland behind the tympanum. It can be seen by reflecting the hinder border of the tympanum forwards and removing the adjacent muscle that lowers the mandible. (See p. 59.)

CHAPTER II.

THE VASCULAR SYSTEM OF THE FROG.

THE vascular system is a closed system of tubes or vessels filled with blood, and ramifying through all parts of the body: its main parts are: (1) the **heart,** which by its contractions is continually driving the blood round and round the system of vessels: (2) the **arteries,** which are the vessels taking the blood from the heart to all parts of the body: (3) the **veins,** which carry the blood from those parts back to the heart: and (4) the **capillaries,** a system of very small vessels connecting the arteries and veins together.

A. The Heart.

Pin down the frog on its back under water and open the body cavity as before, taking special care to preserve the anterior abdominal vein. (See p. 18.) *In freeing the pectoral girdle from the underlying muscles take care not to injure the neighbouring bloodvessels.*

Open the pericardial cavity and dissect the pericardium from the heart and the roots of the great vessels, examine and draw the heart in situ, showing its several divisions.

1. The divisions of the heart.

i. **The auricles** form the anterior and dorsal division of the heart: they are thin-walled and appear dark in colour owing to the blood being seen through their walls. On close examination the division into right and left auricles can be seen.

ii. **The ventricle** is posterior to the auricles: it is paler in colour owing to the greater thickness of its walls; and is conical in shape, with the apex pointing backwards.

iii. **The truncus arteriosus** is a cylindrical body arising from the right anterior border of the ventricle, and running obliquely forwards across the auricles.

Lift up the ventricle and turn its apex forwards so as to expose the sinus venosus.

iv. **The sinus venosus** is a thin-walled sac, lying dorsal to the ventricle and behind the auricles; it receives the three large **venæ cavæ.**

2. **The pulsation of the heart.**

a. Note that the contractions of the heart continue some time after the frog has been killed, or even after the heart is completely removed from the body.

b. Note the character of the heart's pulsations: a regularly alternating series of contractions and dilatations.

c. Note further that in each contraction or systole of the heart all four divisions of the heart contract, but not simultaneously. The sinus venosus contracts first, then the two auricles, then the ventricle, and finally the truncus arteriosus.

B. The Veins.

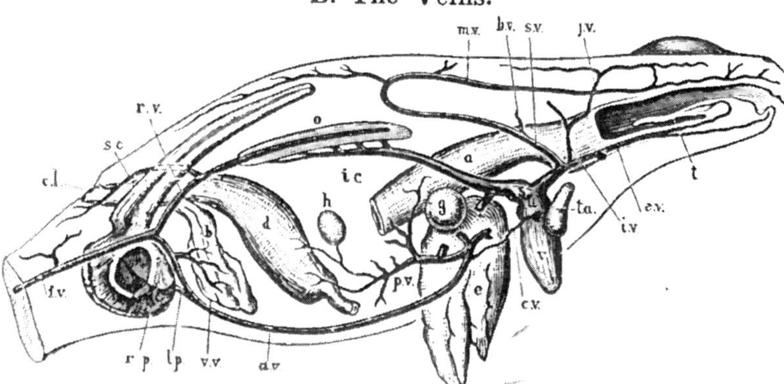

FIG. 4.—Diagrammatic figure of the venous system of the frog, from the right side.

a, stomach; *a.v*, anterior abdominal vein; *b*, bladder; *b.v*, brachial vein; *c.l*, cloacal aperture; *c.v*, cardiac vein; *d*, large intestine; *e*, liver; *e.v*, external jugular vein; *f.v*, femoral vein; *g*, gall-bladder; *h*, spleen; *i.c*, posterior vena cava; *i.v*, innominate vein; *j.v*, internal jugular vein; *l.p*, left pelvic vein; *m.v*, musculo-cutaneous vein; *o*, kidney; *p.v*, hepatic portal vein; *r.p*, right pelvic vein; *r.v*, right renal portal vein; *s*, sinus venosus; *s.c*, sciatic vein; *s.v*, subclavian vein; *t*, tongue; *t.a*, truncus arteriosus; *u*, right auricle; *v*, ventricle; *v.v*, vesical veins.

The veins should be dissected before the arteries, because, as a rule, they lie nearer the surface and are therefore met with first. The veins are further distinguished from the arteries by their larger size and darker colour, due to the blood being seen more clearly through their thinner walls, and to the presence of pigment on them.

Dissect from the ventral surface. In cleaning a vein take hold with the forceps, not of the vein itself but of the tissue surrounding it; and take especial care not to prick the vein, as by doing so you allow the blood to escape and obscure the dissection, and also render the vein itself difficult to see owing to the loss of colour. Always dissect along and not across a bloodvessel, and pin out the parts so as to stretch it slightly.

I. Veins opening into the Sinus Venosus.

a. **The right anterior vena cava** is a large vein opening into the right side of the sinus venosus, and returning to it the blood from the right side of the head and body, and from the right fore limb. It is formed by the union of three veins.

1. The **external jugular vein** is formed by
   i. The **lingual vein**, from the floor of the mouth and the tongue.
   ii. The **mandibular vein**, from the margin of the lower jaw.
   In close connection with the ventral surface of each external jugular vein is a small round vascular body, the **thyroid gland.**

2. The **innominate vein** is formed by
   i. The **internal jugular vein**, returning blood from the interior of the skull, which it leaves by an aperture at the posterior border of the orbit.
   ii. The **subscapular vein**, a small vein from the back of the arm and shoulder.

3. The **subclavian vein**, the largest of the three, is formed by
   i. The **brachial vein**, from the fore limb.
   ii. The **musculo-cutaneous vein**: a very large vein returning blood from the skin and

muscles of the side and back of the body, and of the head as far forwards as the nose.

b. **The left anterior vena cava** corresponds in its course and branches to the right one.

c. **The posterior vena cava** is a median vein which, commencing between the kidneys, runs forward, dorsally to the liver, to open into the posterior end of the sinus venosus. It returns to the heart the blood from the liver and from the kidneys, and indirectly from other viscera and from the hind limbs. It receives the following veins:

i. The **right and left hepatic veins,** from the liver: these open into the posterior vena cava just before it joins the sinus venosus.

ii. The **renal veins,** from the kidneys: of these there are four or five on each side, which open into, or rather form by their union, the posterior vena cava. The most anterior of these receive the veins from the fat bodies.

iii. The **ovarian veins** (in the female), or **spermatic veins** (in the male); returning blood from the ovaries or testes. They are usually four or five in number on each side, and open into the posterior vena cava between the renal veins.

II. Vein opening into the Left Auricle.

a. **The pulmonary vein** is formed by the union of the right and left pulmonary veins, returning to the heart the blood from the right and left lungs respectively. Each pulmonary vein runs along the inner side of its lung.

III. The Portal Systems.

A portal vein is one which, returning blood from the capillaries of some part, breaks up before reaching the heart into a second set of capillaries within some other organ; these again unite to form a vein which carries the blood to the heart. In the frog there are two portal systems, one supplying the kidneys, and the other the liver.

a. The renal portal system.

Trace back the anterior abdominal vein to the hinder end of the body, where it will be seen to be formed by the union of the two **pelvic veins.** *Follow back the pelvic vein of one side to the base of the hind limb; here it will be seen to be one of two branches into which the* **femoral vein,** *the large vein returning blood from the hind limb, divides. The other branch of the femoral vein is the* **renal portal vein,** *which is to be followed to the outer side of the kidney.*

1. The **right renal portal vein** is the dorsal branch of the right femoral vein: it runs forwards along the outer side of the kidney and ends in numerous branches in its substance. It receives the following branches:
   - i. The **right sciatic vein,** from the muscles and skin of the back of the thigh, joins the renal portal vein close to its commencement, before it reaches the kidney.
   - ii. The **right dorso-lumbar veins** are small veins from the dorsal wall of the body, and, in the female, from the oviduct: they join the renal portal vein opposite the kidney.
2. The **left renal portal vein** corresponds in its course and branches to the right vein.

b. The hepatic portal system.

This is formed partly by the anterior abdominal vein, which brings to the liver blood from the hind limbs; and partly by veins returning blood from the alimentary canal.

1. The **anterior abdominal vein** is a median vein formed by the union of the two pelvic veins, the ventral branches of the femoral veins. It runs forwards along the middle line of the ventral body-wall to the level of the liver, where it leaves the body-wall and divides into right and left branches, which enter the right and left lobes of the liver respectively. During its course it receives the following veins:
   - i. **Vesical veins,** from the bladder.
   - ii. **Parietal veins,** from the ventral body-wall.

iii. A **cardiac vein**, from a network of vessels on the truncus arteriosus.

2. The **hepatic portal vein** is a wide vein which runs in the mesentery and joins the anterior abdominal vein at its point of division into right and left branches; giving off, before doing so, a branch to the left lobe of the liver. It carries to the liver the blood from the walls of the alimentary canal, and is formed by the union of the following veins:

   i. The **gastric vein**, from the stomach.

   ii. **Intestinal veins**, from the whole length of the intestine, both small and large.

   iii. The **splenic vein**, from the spleen: this usually joins one of the intestinal veins.

C. The Arteries.

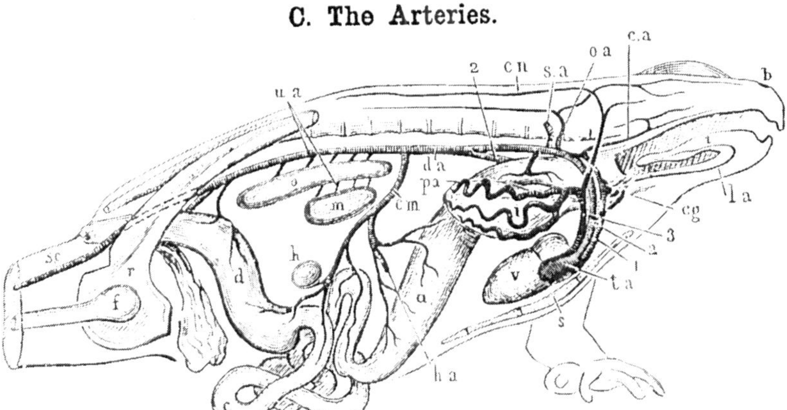

FIG. 5.—Diagrammatic figure of the arterial system of the male frog, from the right side.

a, stomach; *b*, nostril; *c*, small intestine; *c.a*, carotid artery; *c.g*, carotid gland; *c.m*, cœliaco mesenteric artery; *c.n*, cutaneous artery; *d*, large intestine; *d.a*, dorsal aorta; *f*, femur; *h*, spleen; *h.a*, hepatic artery; *i*, right lung; *l.a*, lingual artery; *m*, testis; *o*, kidney; *o.a*, occipito-vertebral artery; *p.a*, pulmonary artery; *r*, pelvic girdle; *s*, sternum; *s.a*, subclavian artery; *s.c*, sciatic artery; *t*, tongue; *t.a*, truncus arteriosus; *u.a*, urinogenital arteries; *v*, ventricle; 1, carotid arch; 2, systemic arch; 3, pulmo-cutaneous arch.

Dissect as for the veins. Pass a roll of paper or plugs of cotton-wool down the œsophagus, so as to distend it and stretch the aortic arches. Clean carefully the aortic arches, commencing at the truncus arteriosus; and follow the several arteries to their distribution, removing the veins and other structures which overlie them. Note the division of the truncus arteriosus in front into right and left branches, each of which again divides into three aortic arches—the **carotid arch,** *the* **systemic arch,** *and the* **pulmo-cutaneous arch.**

I. The Carotid Arch is the most anterior of the three arches: it runs round the side of the œsophagus, and is connected dorsally with the second or **systemic arch**; its chief branches are as follows:

1. The **external carotid artery** is a small artery supplying the tongue. Immediately beyond the origin of the lingual artery the carotid arch presents a small spongy swelling, the **carotid gland.**
2. The **internal carotid artery** runs round the side of the œsophagus to its dorsal surface: it is connected with the systemic arch by a short branch, the ductus Botalli, which in the adult frog is usually impervious; and then turns forwards beneath the base of the skull, dividing in front into the two following vessels:
   i. The **palatine artery,** supplying the roof and sides of the buccal cavity, and the orbit.
   ii. The **cerebral artery,** which enters the skull and supplies the brain.

II. The Systemic Arch, the middle arch of the three, runs somewhat obliquely round the œsophagus to the dorsal surface, and unites with its fellow of the opposite side about the level of the anterior ends of the kidneys. From this point the right arch gives off the main artery to the viscera, the left arch being the chief source of the **dorsal aorta.** Near the level of the posterior ends of the kidneys the aorta divides into the two **iliac arteries.** The branches of the systemic arch are as follows:

a. Branches given off before the union of the two arches.

1. The **laryngeal artery** is a small branch arising from the inner side of the systemic arch near its origin from the truncus arteriosus, and supplying the larynx.
2. The **œsophageal arteries** are one or two branches arising

from the upper part of the arch and entering the dorsal wall of the œsophagus.

3. The **occipito-vertebral artery** is a short branch arising from the dorsal part of the arch: it runs upwards immediately in front of the transverse process of the second vertebra, and divides into two:
   i. The **occipital artery**: which runs forwards, supplying the side of the head and jaws.
   ii. The **vertebral artery**: a large artery which runs back alongside of and above the vertebral column, and gives branches to the muscles of the body-wall and to the spinal cord.
4. The **subclavian artery**: arises from the arch immediately behind the occipito-vertebral artery, and runs outwards, supplying the shoulder and fore-limb.

Branches given off after the union of the two arches to form the dorsal aorta.

1. The **cœliaco-mesenteric artery** is a large median artery arising immediately beyond the point of union of the two arches, or sometimes from the left arch just before the union, and supplying the stomach and intestines. Its branches are as follows:
   i. The **cœliac artery**: which divides into
      α. The **gastric artery**, supplying the stomach.
      β. The **hepatic artery**, supplying the liver and gall-bladder.
   ii. The **mesenteric artery**: which divides into
      α. The **anterior mesenteric artery**, supplying the proximal part of the intestine.
      β. The **posterior mesenteric artery**, supplying the distal part of the intestine.
      γ. The **splenic artery**, supplying the spleen.
2. The **urinogenital arteries** are four to six small arteries which arise from the ventral surface of the aorta between the kidneys, and immediately divide into right and left branches, supplying the kidneys, the reproductive organs and ducts, and the fat bodies.
3. The **lumbar arteries** are small paired lateral branches supplying the body-walls.

4. The **hæmorrhoidal artery** is a small median artery arising from the hinder end of the aorta, and supplying the large intestine.

c. Branches formed by the division of the aorta.

1. The **iliac arteries** are the two large arteries formed by the division of the aorta, and supplying the hind-limbs. Each gives off a **hypogastric artery**, which supplies the bladder, giving **epigastric** branches to the ventral body-wall, and then continues as the **sciatic artery** down the leg, giving off branches to the muscles and skin of the thigh, and dividing at the knee into **peroneal** and **tibial arteries** supplying the leg and foot.

III. The Pulmo-cutaneous Arch is the hindmost of the three aortic arches: it divides about the level of the carotid gland into the following branches:

1. The **cutaneous artery** is a large artery which at first runs forwards and upwards and then turns backwards, supplying the skin of the back along the whole length of the body, and sending smaller branches to the sides of the head and to the skin of the ventral surface.
2. The **pulmonary artery** runs with somewhat sinuous course along the outer side of the whole length of the lung, giving off branches into its walls.

D. The Structure of the Heart.

Having completed the dissection of the bloodvessels, cut them across, about half an inch from the heart; remove the heart completely, and dissect it carefully under water. It is well to cut the vessels of unequal lengths on the two sides, as this will facilitate the recognition of the sides of the heart during the dissection.

Place the heart at first with the dorsal surface upwards.

1. **The Sinus Venosus** (Fig. 4, p. 25) is a thin-walled sac on the dorsal surface of the heart; it is triangular in shape, with the apex directed backwards. Into its anterior angles the right and left anterior venæ cavæ open, and into its posterior angle or apex the posterior vena cava.

Cut away with scissors the dorsal wall of the sinus venosus so as to expose its cavity: wash out any contained blood.

The **sinu-auricular aperture** (Fig. 6, SV) leading from the sinous venosus to the right auricle, is a transversely oval opening, guarded by imperfect anterior and posterior valves, in the ventral wall of the sinus venosus, close to its anterior end, and very nearly in the median plane.

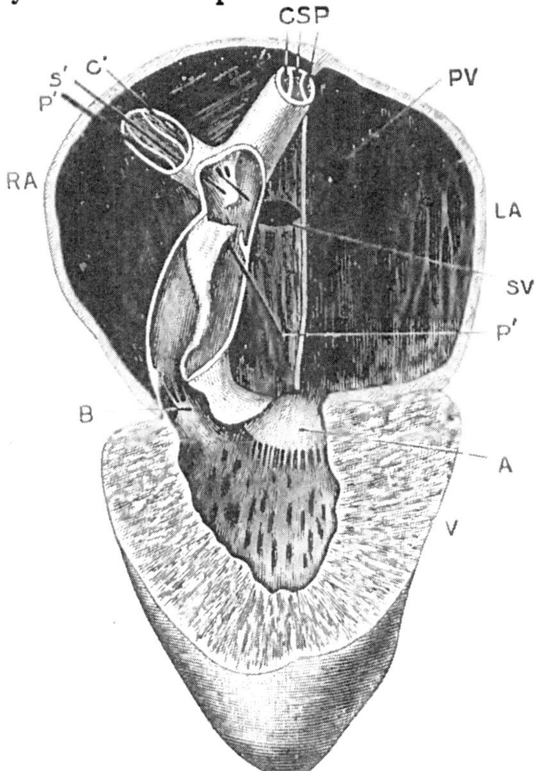

FIG. 6.—The frog's heart seen from the ventral surface, and dissected so as to show its structure. The ventral walls of the truncus arteriosus, and of the auricles and ventricle have been removed. (From a drawing by Dr. Hurst.)

A, auriculo-ventricular aperture and one of its valves; B, aperture leading from ventricle to truncus arteriosus, with one of its valves; C, left carotid arch; C', style passed down right carotid arch into the truncus arteriosus; LA, left auricle; P, left pulmo-cutaneous arch; P'P', style, passed down right pulmo-cutaneous arch into the truncus arteriosus; PV, opening of pulmonary vein into left auricle: RA, right auricle; S, left systemic arch; S', style passed down right systemic arch into the truncus arteriosus; SV, opening from sinus venosus into right auricle; V, ventricle.

2. **The Auricles.** *Turn the heart over, with its ventral surface upwards. Cut away the ventral wall of both auricles with fine scissors, taking care not to damage the truncus arteriosus which lies across the right auricle. Wash out the blood from the auricles.*

a. The **right auricle** (Fig. 6, RA) is the larger of the two. It has thin walls, thickened by muscular strands which form interlacing reticular ridges on its inner surface. In the dorsal wall of the auricle, very near the median plane of the heart, is the aperture from the sinus venosus already described (Fig. 6, SV).

b. The **left auricle** (Fig. 6, LA) is smaller, sometimes much smaller, than the right auricle, which it resembles in the structure of its walls. In its dorsal wall, very close to the sinu-auricular aperture, is the opening of the pulmonary vein (Fig. 6, PV).

c. The **interauricular septum** is the thin partition between the right and left auricles. It is much thinner than the walls of the auricles, and is placed somewhat obliquely, the left auricle lying rather more dorsally than the right. The septum ends with a free posterior edge, opposite the auriculo-ventricular aperture.

Cut away with scissors the ventral wall of the ventricle, taking care not to damage the truncus arteriosus.

3. **The Ventricle** (Fig. 6, V) is conical in shape with the apex backwards, and has a small central cavity, with thick spongy walls. The spongy character is due to great development of a reticulum of interlacing muscular strands similar to those of the auricles: the true outer wall of the ventricle is no thicker than that of the auricles, and the meshes of the spongework are really part of the cavity of the ventricle, and are filled with blood.

The **auriculo-ventricular aperture** lies at the base of the ventricle, and rather to the left side. It is guarded by valves (Fig. 6, A) which hang into the ventricle, and are tied down at their edges by fine tendinous threads; and it is divided by the free lower edge of the interauricular septum into right and left divisions, admitting blood from the right and left auricles respectively.

Cut away carefully, with fine scissors, the ventral wall of the truncus arteriosus so as to expose its cavity and the contained valves.

4. The **Truncus Arteriosus** consists of two parts; a proximal part or **pylangium**, which is a single vessel arising from the ventricle; and a distal part or **synangium**, which consists of the basal parts of the aortic arches closely united together.

a. The **pylangium** (Fig. 6) is a short tube arising from the right-hand ventral corner of the anterior end of the ventricle: it has thick muscular walls and is widest about the middle of its length.

The opening from the ventricle to the pylangium (Fig. 6, B) is guarded by three semilunar pocket valves.

The opening from the pylangium to the synangium is also guarded by three semilunar valves which are of very unequal size, a large right one, a small left one, and a still smaller dorsal valve.

The **spiral valve** is a longitudinal ridge, projecting into the cavity of the pylangium: it commences at the left side of the ventricular aperture and runs forwards somewhat spirally along the dorsal wall of the pylangium to its anterior end, where it fuses with the large right valve of the three between the pylangium and the synangium. The ventral edge of the spiral valve is free and rounded, and the valve is much wider at its anterior than at its posterior end.

b. The **synangium** is the distal part of the truncus arteriosus. In its dorsal wall immediately beyond the valves separating it from the pylangium, is an aperture (Fig. 6, P′), leading to the right and left pulmo-cutaneous arches, P,P′. Beyond this the synangium contains a wide cavity continued right and left into the two systemic arches—S,S′. The cavity is partially divided by a vertical tongue-like projection from its dorsal wall: on the ventral surface of this tongue are two small openings, very close together, which lead into the right and left carotid arches, C,C′, both of which arise from the right side.

Cut across the aortic arches, just beyond the division of the truncus into right and left branches, and note that though each branch is apparently a single vessel its cavity is really divided into three vessels corresponding to the three aortic arches. Pass bristles down these aortic arches, and note the points at which they severally open into the truncus arteriosus.

E. The Lymphatic System.

The lymphatic system forms an accessory part of the vascular system. Its main divisions are as follows:

1. The **lymphatic vessels** are a series of thin-walled tubes, very variable in diameter and irregular in shape, which traverse all the parts and organs of the body and are in free communication with the veins. They are of small size, and can only be recognised with the microscope.
2. The **lymph sacs** are large irregular spaces communicating with the lymphatic vessels. The most important are the following:
    a. The **subcutaneous lymph sacs** are the large cavities between the skin and the muscles, which have already been seen when removing the skin.. They are separated from one another by narrow septa of connective tissue, which bind the skin to the underlying body-wall.

    b. The **abdominal lymph sacs** are the large spaces along the dorsal surface of the body-cavity, ventral to the kidneys, and between the peritoneum and the body-walls. (See Fig. 2, p. 19.) The body-cavity itself also communicates with the lymphatic system through small openings or stomata in the peritoneum.
3. The **lymph hearts** are two pairs of small globular contractile sacs placed at points where the lymphatic vessels communicate with the veins. They are quite transparent.
    a. The **anterior lymph hearts** lie immediately behind the transverse processes of the third

vertebra, and beneath the shoulder girdle: they open into the subscapular veins.

b. The **posterior lymph hearts** lie at the sides of the urostyle, close to its hinder end. They communicate by short vessels with the femoral veins. Their pulsations can easily be seen in a pithed frog.

4. The **spleen** has been already referred to (p. 23).

F. Microscopic Examination of Blood.

I. Frog's Blood.

1. **Normal.**

Place on a slide a small drop of blood from the heart of a frog; dilute it with a drop of normal salt solution (0·75 *per cent.*); *put on a thin cover-glass, and run a ring of oil round the edge to prevent evaporation: examine with the high power.*

Blood consists of a colourless fluid, the **liquor sanguinis** or **plasma**, in which float the **blood corpuscles.** These corpuscles are of two kinds.

i. **Red corpuscles.** These are very numerous, pale red or yellowish red in colour, and of a flattened oval shape, with rounded edges and a central bulging, the **nucleus.** The flattened shape is best seen when a corpuscle turns edgeways. They measure 0·0235 mm. in length by 0·0145 mm. in width; or about $\frac{1}{1100} \times \frac{1}{1800}$ of an inch.

ii. **White corpuscles.** These are much fewer in number and of smaller size: they are colourless, granular, subspherical in shape, and exhibit "amœboid" movements. *Sketch one half a dozen times at intervals of half a minute.*

In order to find the white corpuscles, focus the red corpuscles as sharply as possible; then on turning the screw of the fine adjustment slightly, so as to throw them just out of focus, the white corpuscles stand out as blurred white points (owing to their refracting light more strongly); when thus found, focus downwards again, and study them.

2. **Action of acetic acid on blood.**

Place a fresh drop of blood on a clean slide: add a drop of acetic acid: cover, and examine with the high power: note the changes produced.

i. **Red corpuscles:** the nuclei become much more apparent than before, and the red colour disappears.

ii. **White corpuscles:** become clearer, and show nuclei, sometimes more than one in a single corpuscle.

II. Human Blood.

1. **Normal.**

Prick the tip of your finger, and place a small drop of the blood on a slide: add a drop of normal salt solution, cover, and examine as before. Note the following points:

i. **Red corpuscles.** These, which are much smaller than in frog's blood, are in the form of circular biconcave discs with rounded edges but no nuclei. They have a tendency to run together into rouleaux, like piles of coins. Their average diameter is 0·008 mm., or about $\frac{1}{3200}$ of an inch.

ii. **White corpuscles.** These are very similar to those of the frog: they are slightly larger than the red corpuscles, averaging about 0·01 mm., or $\frac{1}{2500}$ of an inch in diameter: their amœboid movements are not well seen unless the slide is warmed.

2. **Action of acetic acid.**

Treat with acetic acid as before: note that, unlike the frog's blood, no nuclei are visible in the red corpuscles.

G. Circulation of the Blood in the Web of a Frog's Foot.

The web uniting the toes of the frog's foot is so thin and transparent, that with the microscope the blood in it can readily be seen coursing along the capillaries.

Examine a frog prepared to show the circulation in the web of the foot. Note the following points:

1. With a low power.

a. The irregularly **branched pigment cells** to which the colour of the frog's skin is due.

b. The fine meshwork of bloodvessels along which the blood can be seen flowing. These bloodvessels are of three kinds.

i. The **arteries**, carrying blood to the web, are distinguished by the fact that when they divide, the direction of flow of the blood is from the larger trunk to its branches.

ii. The **capillaries** form a close network of very small and thin-walled vessels, along which the blood flows from the arteries to the veins.

iii. The **veins**, carrying the blood away from the web back towards the heart, are distinguished from the arteries by the fact that the blood in them flows from smaller to larger vessels.

2. With a high power: *note the following points:*

a. The walls of the arteries and veins are much thicker than those of the capillaries, which latter are often difficult to see.

b. The white corpuscles have a marked tendency to creep along the sides of the vessels, while the red corpuscles rush far more rapidly along the middle of the stream: this is seen best in the small arteries.

c. The variation in calibre of the small arteries and capillaries: whilst under observation an artery or capillary may be seen to change its size to a considerable extent.

d. The indefinite character of the capillary circulation. Owing to changes of size in adjacent vessels, the direction of flow of the blood in a given capillary may become reversed.

e. The elasticity of the red corpuscles: seen best when they are turning the corners of the capillary network.

f. The tendency of the white corpuscles to migrate through the walls of the capillaries into the tissues outside. This is much increased by the application of some irritant substance, such as a drop of weak acid, to the web.

CHAPTER III.

THE SKELETON OF THE FROG.

THE skeleton, which forms the hard internal parts of the frog, is composed partly of cartilage and partly of bone. It forms a framework giving definite shape to the body, and precision to the movements; and serves also to protect from injury some of the more important and delicate organs, notably the central nervous system, the sense organs and the heart. In the early stages of its development the skeleton consists entirely of cartilage; in the adult this primary cartilaginous skeleton is replaced to a greater or less extent by bone. Bone may also be developed in places where there was no pre-existing cartilage, and is then called *membrane-bone*, in contradistinction to the former kind, or **cartilage-bone**, which replaces cartilage. Membrane-bones arise in the first instance as ossifications in the dermis or deeper layer of the skin: in many fish they retain this primitive position, but in the frog and most higher vertebrates they sink below the skin and graft themselves on to the more deeply placed cartilaginous skeleton. Cartilage may also become calcified, *i.e.*, have calcareous salts deposited in its matrix, without in any way taking on the character of true bone.

The skeleton may conveniently be divided into (1) the **axial portion**, including the skull and the vertebral column: and (2) the **appendicular portion**, including the limbs, and the limb-girdles which attach them to the body.

Examine the prepared skeletons, and make careful drawings to scale of the several parts. In your drawings colour the cartilage blue, the cartilage-bones yellow, and the membrane-bones white or red. Prepare skeletons for yourself by dipping the parts in hot water, and carefully brushing away the soft tissues until the skeleton is clean.

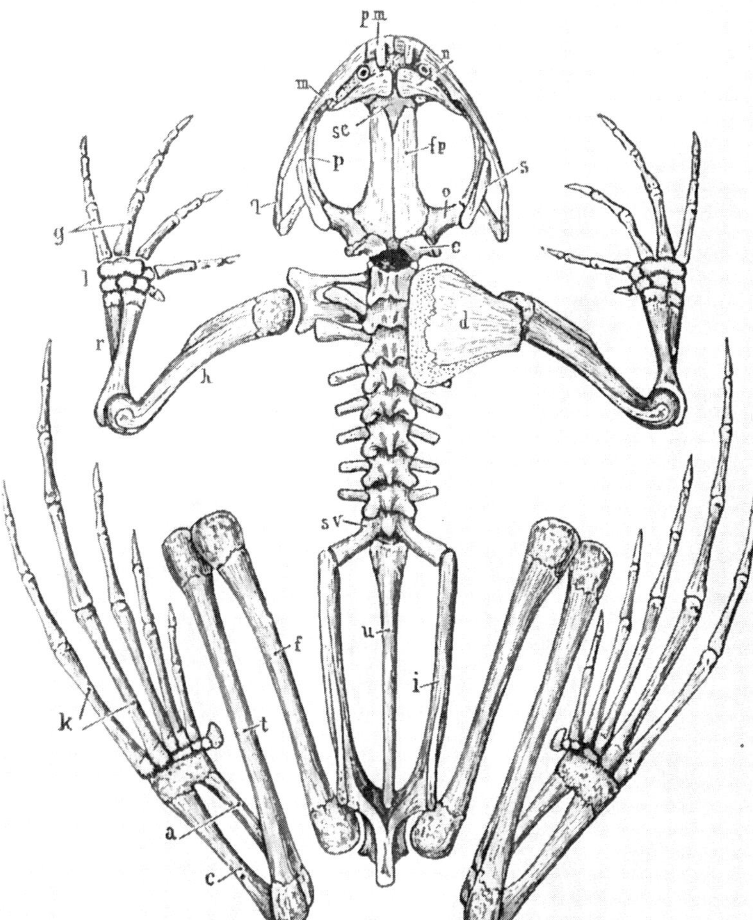

FIG. 7.—The skeleton of the frog, seen from the dorsal surface; the left suprascapular and scapular have been removed.

a, astragalus; *c*, calcaneum; *d*, suprascapular; *e*, exoccipital; *f*, femur; *fp*, frontoparietal; *g*, metacarpals; *h*, humerus; *i*, ilium; *k*, metatarsals; *l*, carpus; *m*, maxilla; *n*, nasal; *o*, pro-otic; *p*, pterygoid; *pm*, premaxilla; *q*, "quadratojugal"; *r*, radio-ulna; *s*, squamosal; *se*, sphenethmoid; *s.v*, sacral vertebra; *t*, tibio-fibula; *u*, urostyle.

A. The Axial Skeleton.

I. The Vertebral Column or "backbone."

This is a bony tube which surrounds and protects the spinal cord; it consists of an anterior part, which is divided transversely into nine rings or **vertebræ**, and a posterior unsegmented portion of about equal length—the **urostyle.** At the sides of the tube, between the successive vertebræ, are the **intervertebral foramina** through which the nerves pass out from the spinal cord to the various parts of the body.

a. **Structure of a vertebra.** *Examine one of the vertebræ, say the third, more closely: draw it, showing the following points:*

i. The **vertebra** is a bony ring; the spinal cord lying during life in the central **neural canal.**

ii. The **centrum** or **body** is the thickened ventral portion of the ring: it articulates with the centra of the vertebræ in front of and behind it; and forms the floor of the neural canal.

iii. The **neural arch** consists of the lateral and dorsal portions of the ring; and forms the sides and roof of the neural canal.

iv. The **spinous process** or **neural spine** is a small blunt median process, projecting upwards and backwards from the top of the neural arch.

v. The **transverse processes** are a pair of large processes projecting horizontally outwards from the point of union of centrum and neural arch.

vi. The **articular processes** or **zygapophyses**, on the anterior and posterior borders of the neural arch, articulate with corresponding processes on the vertebræ in front and behind, and so serve to link the vertebræ together.

α. The **anterior articular processes**, or **prezygapophyses**, face upwards and slightly inwards.

β. The **posterior articular processes**, or **postzygapophyses**, face downwards and slightly outwards.

b. **Special vertebræ.**

i. The **atlas** or **first vertebra** articulates in front with the posterior end of the skull: it has no transverse processes. Note the large gap on the dorsal surface between the skull and the neural arch of the atlas: through this gap, which is closed by the strong **occipito-atlantal membrane,** the central nervous system is divided and destroyed in the operation of pithing a frog.

ii. The **sacrum,** or ninth vertebra, has very stout backwardly directed transverse processes, which support at their outer ends the pelvic arch.

c. The **urostyle** is the unsegmented posterior portion of the vertebral column. It articulates in front with the body of the sacral vertebra by two surfaces. Along its dorsal surface runs a prominent vertical ridge, highest in front and gradually diminishing posteriorly: the neural canal is continued down the anterior part of this ridge. At the sides of the urostyle, and about the length of a vertebra from its anterior end, are a pair of small holes through which nerves pass out, and which therefore correspond to intervertebral foramina.

II. The Skull.

The skull consists of, (1) an axial portion, the **cranium,** enclosing the brain and forming an anterior continuation of the vertebral column; (2) the **olfactory capsules** and the **auditory capsules,** which are fused with the anterior and posterior ends of the cranium respectively; (3) the bony framework of the jaws; and (4) the **hyoid apparatus.**

In the skull the original cartilage, or chondrocranium, is not so largely replaced by bone as in the vertebral column, large tracts of unossified cartilage persisting in the adult. Besides the **cartilage-bones** the skull is further strengthened by the addition of numerous *membrane-bones.*

1. **The Cranium** is originally an unsegmented cartilaginous tube, whose cavity forms the anterior part of the neural canal, and lodges the brain. The roof of the

tube is imperfect, there being one large **anterior fontanelle,** and two smaller **posterior fontanelles,** which are closed by membrane only. In the cartilage are developed cartilage-bones, and around it membrane-bones.

To study the cranium satisfactorily, the membrane-bones should be stripped from one of the skulls which you have prepared for yourself.

a. Cartilage-bones of cranium.

i. The **exoccipitals** are two irregular bony masses at the sides of the posterior end of the skull. They almost completely surround the **foramen magnum** or entrance to the cranial cavity; and bear on their posterior surfaces the **occipital condyles,** two oval convex processes which articulate with the first vertebra or atlas.

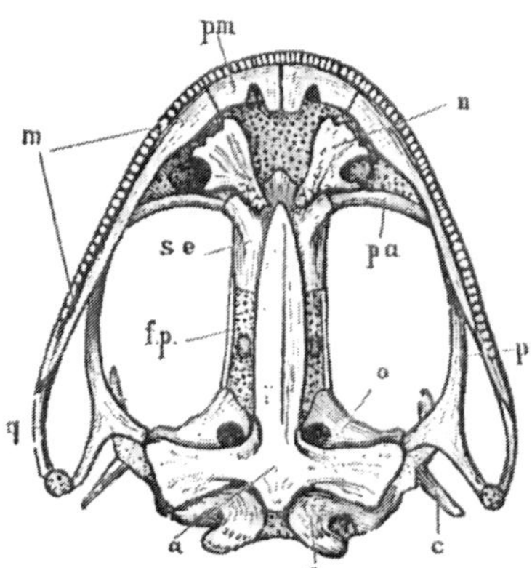

FIG. 8.—The frog's skull, from the ventral surface.
a, parasphenoid; *c*, columella; *e*, exoccipital; *f.p*, frontoparietal; *m*, maxilla; *n*, vomer; *o*, pro-otic; *p*, pterygoid; *pa*, palatine; *pm*, premaxilla; *q*, "quadratojugal"; *se*, sphenethmoid.

ii. The **sphenethmoid** or **girdle-bone** is a bony tube which encircles the anterior end of the cranial cavity, and extends forwards into the olfactory region: in front it is divided by a vertical partition into right and left cavities, in which lie the olfactory sacs.

b *Membrane-bones of cranium.*

i. The *frontoparietals* are two long flat bones on the top of the brain-case, covering the fontanelles, and overlapping the hinder end of the sphenethmoid.

ii. The *parasphenoid* is a ⊥-shaped bone on the ventral surface of the cranium; its lateral processes underlying the auditory capsules.

2. **The Sense Capsules** are cartilaginous and bony capsules which surround and protect the olfactory, optic, and auditory organs; two of them are fused with the cranium so as to form parts of the skull.

a The **auditory capsules** are fused with the sides of the posterior end of the cranium, to which they form wing-like projections: they consist largely of cartilage.

i. **Cartilage-bone of auditory capsules.**

The **pro-otics** are a pair of irregular-shaped bones in the anterior walls of the capsules, and forming also parts of their roof and floor.

b. The **optic capsules** are not fused with the skull. They remain permanently cartilaginous under the name of the sclerotic cartilages of the eyes.

c. The **olfactory capsules** are fused with the anterior end of the cranium, and also with each other. They consist very largely of cartilage, which is produced in front into the **rhinal processes**.

i. **Cartilage-bone of olfactory capsules.** The **spheneth-moid**, as already noticed, extends forwards so as to invade the olfactory region, but does not properly belong to the olfactory capsules.

ii. *Membrane-bones of olfactory capsules.*

a. The *nasals* are two triangular bones on the dorsal surface of the anterior end of the head: the bases of the triangles are turned towards the middle line and meet each other in front, while their posterior ends diverge and enclose, with the anterior ends of the frontoparietals, a diamond-shaped patch in which the sphenethmoid is visible on the dorsal surface of the skull.

β. The *vomers* are two triradiate bones on the ventral surface of the fore part of the skull: each vomer bears in its inner or posterior angle a small group of pointed teeth, and forms the inner boundary of the posterior narial opening of its side.

3. **The Jaws** consist of two cartilaginous arches on each side, maxillary and mandibular, in connection with which **cartilage-bones** and *membrane-bones* are developed. Each arch meets its fellow in the middle line in front; and the maxillary arches, forming the upper jaw, are firmly connected with the cranium by anterior and posterior bony struts.

a. The **Maxillary Arch.**

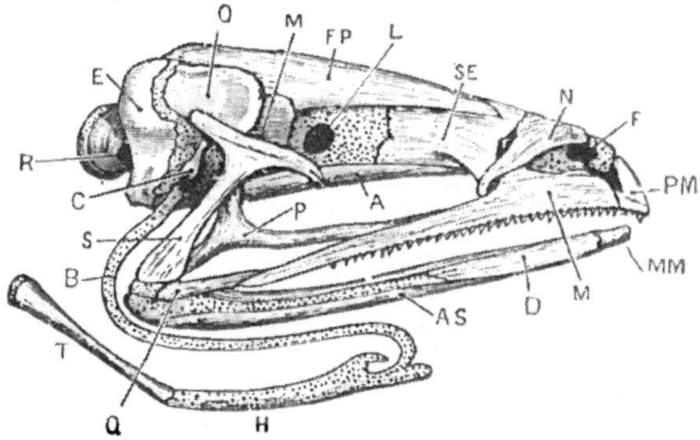

FIG. 9.—The frog's skull from the right side.

A, parasphenoid; AS, angulosplenial; B, anterior cornu of hyoid; C, columella; D, dentary; E, exoccipital; F, nostril; FP, fronto-parietal; H, body of hyoid; L, aperture for exit of optic nerve; M, maxilla; MM, mentomeckelian; M′, aperture for exit of fifth and seventh nerves; N. nasal; O, pro-otic; P. pterygoid; PM, premaxilla; Q, "quadratojugal"; R, aperture for exit of ninth and tenth nerves; S, squamosal; SE, sphenethmoid; T, posterior cornu of hyoid.

i. The **pterygoid*** is a large triradiate bone, the inner limb of which is connected with the auditory capsule; while the posterior limb runs back to the angle of the mouth, and the anterior limb forwards along the upper jaw to the palatine bone.

ii. The **palatine*** is a slender transverse bone, con-

* These, originally cartilage-bones, become more or less ensheathed by membrane-elements in the course of development.

necting the upper jaw with the anterior end of the sphenethmoid.

iii. A short bone forming the posterior part of the outer margin of the upper jaw, is in the position of the *quadratojugal* of other skulls.

iv. The *maxilla* is a long thin bone forming the greater part of the outer margin of the upper jaw: it bears teeth along its whole length which are anchylosed with the bone. It is connected behind with the "quadratojugal"; about the middle of its length with the anterior limb of the pterygoid and with the palatine; and in front with the premaxilla.

v. The *premaxilla* is a small bone which meets its fellow in the middle line in front, and so completes the outer margin of the upper jaw: like the maxilla it bears teeth. It gives off on its dorsal surface a backwardly projecting process which forms part of the inner boundary of the nostril.

vi. The **quadrate cartilage,** which forms the **suspensorium,** *i.e.*, serves to connect the lower jaw with the skull, is a rod of cartilage which is fused above with the auditory capsule, and runs downwards and backwards to the angle of the mouth, where it is connected with the hinder end of the quadratojugal bone. In the adult frog, the quadrate cartilage lies between the squamosal and pterygoid bones, and is almost completely concealed by these.

vii. The *squamosal* is a T-shaped bone, the stem of which is closely applied to the outer surface of the quadrate cartilage. The posterior limb of the T is attached to the outer surface of the auditory capsule, and with the body of the squamosal helps to support the annulus tympanicus.

b. The **Mandibular Arch.** The arch persists in part unossified as **Meckel's cartilage,** which forms the basis of the lower jaw, and is ensheathed by cartilage-bones and membrane-bones.

i. The *angulosplenial* ensheathes the inner and lower surfaces of Meckel's cartilage along the greater part of its length: near its hinder end it is produced upwards into the coronary process.

ii. The *dentary* is a flat bone covering the outer surface of the distal half of Meckel's cartilage, as far forward as the mentomeckelian bone.

iii. The **mentomeckelian** is a small ossification in Meckel's cartilage at the symphysis, *i.e.*, the union of the arches of the two sides at the chin.

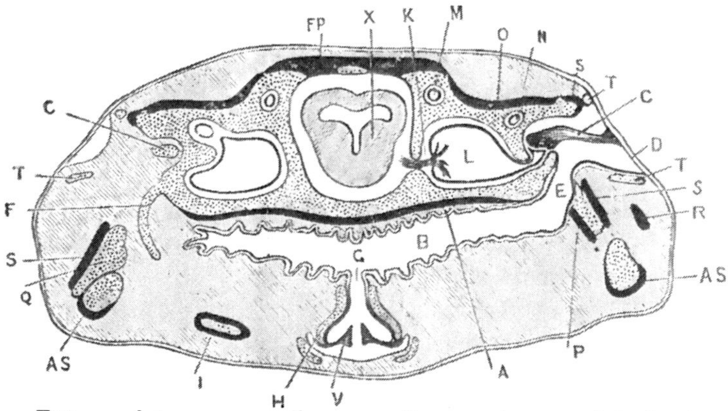

FIG. 10.—A transverse section across the posterior part of the frog's head, to show the position and relations of the auditory organs, Eustachian tubes, and hyoid apparatus. On the right side the section passes through the tympanic cavity and the columella; on the left side through the anterior cornu of the hyoid. The cartilage is dotted, and the bones, except the columella, are represented black.

A, parasphenoid; AS, angulosplenial; B, buccal cavity; C, columella; D, tympanic membrane; E, Eustachian tube; F, anterior cornu of the hyoid; FP, frontoparietal; G, glottis; H, arytenoid cartilage; I, posterior cornu of hyoid; K, auditory nerve; L, vestibule; M, anterior vertical semicircular canal; N, horizontal semicircular canal; O, pro-otic; P, pterygoid; Q, quadrate cartilage; R, "quadratojugal"; S, squamosal; T, annulus tympanicus; V, vocal cord; X, mid-brain.

4. The Hyoid Apparatus (Fig. 9). This consists of the hyoidean arch and the remains of the branchial arches of the two sides, together with a median ventral plate, the **body of the hyoid**, which unites their lower ends together, and lies in the floor of the mouth. The hyoid apparatus consists almost entirely of cartilage.

a. The **hyoid arch.**

i. The **columella** (Figs. 9 and 10) is formed from the top of the hyoid arch: it is a small rod, partly bone and partly cartilage, the inner end of which is inserted into the **fenestra ovalis**, an aperture in the outer wall of the auditory capsule; while the outer end is attached to the tympanic membrane rather above its middle.

ii. The **anterior cornu of the hyoid** (Fig. 9) is a long slender curved rod of cartilage, attached above to the auditory capsule just below the fenestra ovalis, and curving at first backwards and then forwards and downwards to be attached to the anterior outer angle of the body of the hyoid.

b. The **body of the hyoid** is a flat squarish plate of cartilage, formed by the fused ventral ends of the hyoid and branchial arches, and lying in the floor of the mouth: short processes are given off from its angles.

c. The **posterior cornua of the hyoid** are a pair of stout bony processes, diverging from the hinder border of the body of the hyoid.

B. The Appendicular Skeleton.

This comprises the limbs and the limb-girdles. As in the case of the axial skeleton it consists at first entirely of cartilage, which becomes afterwards replaced to a greater or less extent by cartilage-bone. Membrane-bones, *i.e.*, bones developed independently of cartilage, are very rare, the clavicles being the only examples met with in the frog. (Fig. 7, p. 41.)

1. The Pectoral Girdle.

This consists originally of two half rings of cartilage, one on each side of the body, which they encircle a short way behind the head: the dorsal ends of the half rings are attached by ligaments and muscles to the vertebral column, while the ventral ends are united together in the median plane by the **sternum** or " breast bone."

Each half ring bears in the middle of its hinder surface a cup-shaped cavity, which, with the first bone of the fore-limb.

forms the shoulder-joint. The part of the arch above the joint is the **scapular portion**: and the part below, which is divided into anterior and posterior divisions, the **coracoid portion.**

a. The **scapular portion** is divided into two parts.

i. The **suprascapula,** the upper portion, is a thin expanded plate of cartilage overlapping the first four vertebræ: it is partly calcified and partly ossified, though very imperfectly.

ii. The **scapula** is an oblong bony plate, constricted in the middle, and forming the upper half of the **glenoid cavity,** or cavity of the shoulder-joint.

b. The **coracoid portion** forms the lower half of the glenoid cavity: it is divided into anterior and posterior portions, separated by the **coracoid foramen.**

i. The **precoracoid** is a slender horizontal bar of cartilage connecting the anterior edge of the scapula with the sternum.

ii. The *clavicle* is a slender bone, closely applied to the anterior border of the precoracoid; its outer or scapular end is bent forwards almost at a right angle.

iii. The **coracoid** is a stout bone, wider at its inner than its outer end: it connects the posterior edge of the scapula with the sternum.

iv. The **epicoracoids** are a pair of narrow strips of cartilage closely applied to each other, and lying between the ventral ends of the precoracoids and coracoids.

c. The **sternum** lies in the mid-ventral line: it consists originally of two lateral halves, which fuse completely in front and behind, but remain distinct in the median portion. It presents from before backwards the following parts:

i. The **presternum.**

α. The **episternum** is a flat circular plate of cartilage.

β. The **omosternum** is a slender bony rod projecting forwards in front of the clavicles.

ii. The **mesosternum** is a rod of cartilage, ensheathed in bone, projecting backwards behind the coracoids.

iii. The **metasternum** or **xiphisternum** is a broad expanded plate of cartilage at the hinder end of the sternum.

2. **The Fore limb.**

The bones of the fore-limb are all cartilage-bones. With the exception of the small bones of the wrist, they are elongated, with enlarged ends capped with articular cartilage. The enlarged ends or **epiphyses** ossify independently of the shaft of the bone, with which they do not unite until late in life. The end of a bone which, when the limb is extended, is nearer to the body, is called its **proximal** end, the opposite extremity the **distal** end.

a. The **arm.** In the arm there is only a single bone.

i. The **humerus.** The proximal end or **head** is enlarged, and articulates with the glenoid cavity of the pectoral girdle: below the head is the strong **deltoid ridge** extending along the proximal half of the anterior surface. At the distal end is a spheroidal articular surface for the bone of the forearm: and at either side of this a prominent **condylar ridge**, the inner or postaxial one being the larger of the two, especially in the male frog.

b. The **forearm.**

i. The **radio-ulna** corresponds to two bones, **radius** and **ulna,** in other animals: it is single at its proximal end, but in its distal half is imperfectly divided by a groove into anterior or radial, and posterior or ulnar portions. Its proximal end is hollowed out to articulate with the lower end of the humerus, and so form the elbow-joint, behind which it projects backwards as the **olecranon process.**

c. The **wrist** consists of six small **carpal** bones arranged in two rows, proximal and distal, each row having three bones.

d. The **hand** has four complete digits and a rudimentary **pollex** or thumb. Each digit consists of a proximal **metacarpal** bone, beyond which are a variable number of **phalanges**.

i. The **pollex**, the anterior or preaxial digit, consists simply of a small metacarpal bone.

ii. The first complete digit, corresponding to the fore-finger of man, consists of a metacarpal and two phalanges.

iii. The second digit, corresponding to the middle finger of man, consists of a metacarpal and two phalanges.

iv. The third, corresponding to the ring-finger of man, consists of a metacarpal and three phalanges.

v. The postaxial digit, corresponding to the little finger of man, consists of a metacarpal and three phalanges.

3. The Pelvic Girdle.

This consists primitively, like the pectoral girdle, of a couple of half-rings of cartilage, fused together below and attached above to the tips of the transverse processes of the sacrum. In the adult frog the girdle is placed very obliquely, so as to be nearly parallel with the vertebral column instead of at right angles to it.

Each half presents on its outer aspect a cup-shaped cavity—the **acetabulum**—forming, with the thigh bone, the hip-joint: we accordingly distinguish an **iliac portion** above the acetabulum and an **ischio-pubic portion** below it, corresponding respectively to the scapular and coracoid divisions of the pectoral girdle.

i. The **ilium** forms the anterior and upper half of the acetabulum, and extends forwards as an elongated laterally compressed bar, which is attached in front to the transverse process of the sacrum and bears along its dorsal surface a prominent vertical ridge of bone, the **iliac crest**, ending behind in an abrupt vertical border. Posteriorly the two ilia meet each other and are united together in the median plane to form the **iliac symphysis**.

ii. The **pubes** consists entirely of cartilage: it forms the anterior portion of the ventral division of the girdle, and therefore corresponds to the pre-coracoid in the pectoral girdle. The two pubes are completely fused together in the median plane, and form only a very small portion, about one-sixth, of the acetabular cavities.

iii. The **ischium** is the posterior portion of the ventral division and corresponds therefore to the coracoid in the shoulder girdle. It forms the posterior third of the acetabulum. The two ischia are completely fused together in the median plane.

4. The Hind-limb.

The bones have the same general characters as those of the fore-limb, to which they correspond very closely.

a. The **thigh**.

i. The **femur** is a long slender bone, expanded at both ends, and curved slightly in a sigmoid manner. The proximal end or head is spheroidal, and fits into the acetabulum to form the hip-joint: the distal end is somewhat expanded laterally.

b. The **leg**.

i. The **os cruris** or **tibio-fibula** is a single bone, rather longer than the femur, slightly curved, and expanded laterally at both ends. It presents along the greater part of its length a groove indicating its correspondence with two bones, **tibia** and **fibula**, which in man and many other animals remain distinct from each other.

c. The **ankle**, corresponding to the wrist in the fore-limb, consists of two rows of **tarsal** bones.

α. The **proximal row of tarsal bones** consists of two elongated bones united together at both ends, but widely separated in the middle.

i. The **astragalus** is on the preaxial or tibial side.

ii. The **calcaneum** is on the postaxial or fibular side, and is the larger of the two bones.

β. The **distal row of tarsal bones** consists of two very small bones.

d. The **foot** has five complete digits, and a supernumerary toe as well. Each digit consists of a proximal **metatarsal** bone, beyond which are a variable number of **phalanges.**

i. The **hallux** or preaxial digit, corresponding to the great toe of man, is the smallest of the series. It consists of a metatarsal and two phalanges.

On the inner side of the hallux is the **calcar,** supposed to be an additional or supernumerary toe : it may have one or two joints in addition to a short metatarsal.

ii. The second toe consists of a metatarsal and two phalanges.

iii. The third consists of a metatarsal and three phalanges.

iv. The fourth, the longest of the five, consists of a metatarsal and four phalanges.

v. The postaxial digit, corresponding to the little toe in man, consists of a metatarsal and three phalanges.

CHAPTER IV.

THE MUSCULAR SYSTEM OF THE FROG.

THE muscles, or flesh, are the direct means by which the various movements of the body and of its several parts are brought about. A muscle consists of a fleshy belly, which is usually attached at each end by means of tendons to some hard part, very commonly to bone. Motion is effected by the muscle **contracting**, *i.e.*, shortening, and so bringing its two ends, and consequently the parts to which the ends are attached, nearer together. Of the two attachments of a muscle one is usually to a more fixed and central part, the other to a more movable and peripheral part: the former attachment is called the **origin** of the muscle, the latter its **insertion**.

Muscles are of two kinds: (1) **voluntary muscles**, *i.e.*, those which are under the control of the will, as the muscles of the arm: and (2) **involuntary muscles**, *i.e.*, those over which the will has no direct control, as the muscles of the heart and bloodvessels, or of the alimentary canal.

Voluntary muscles, which are the only ones dealt with in this chapter, are usually attached at both ends to bone; but one or other end, or both, may be attached to **aponeuroses**, strong connective tissue membranes which ensheath the muscles and other parts, and separate them from one another.

For the dissection of the muscles, take a frog that has been in spirit for a day or more. When cleaning a muscle be careful to put it on the stretch, and to dissect along, and not across its fibres: define the origin and insertion of the muscle very clearly, and test its action by pulling it gently with the forceps in the direction of its fibres. Always have the skeleton in front of you so as to see accurately the origins and insertions of the muscles. In the following description some of the smaller muscles, especially in the head, are omitted.

A. Muscles of the Trunk.

1. Muscles of the ventral body-wall.

Pin out the frog on its back, remove the skin, and clean the muscles.

i. The **rectus abdominis** runs longitudinally along the mid-ventral wall, the muscles of the two sides being separated from each other in the median plane by the **linea alba**, a longitudinal band of connective tissue, immediately dorsal to which lies the anterior abdominal vein. Each **rectus** muscle is divided into bellies by five transverse tendinous intersections.

The muscle arises from the pubes, runs forwards and is inserted into the dorsal surface of the sternum and coracoid.

ii. The **pectoralis** is a large fan-shaped muscle, consisting of a thoracic portion, which arises from the whole length of the ventral surface of the sternum ; and an abdominal portion, arising from the aponeurosis along the outer side of the **rectus abdominis** almost as far back as the pubes. From this extensive origin the fibres converge to the deltoid ridge of the humerus, into which they are inserted, the line of insertion extending down almost to the elbow.

iii. The **obliquus externus** is a thin sheet of muscle which arises from the aponeurosis of the back, a short distance from the vertebral column, and covers the whole of the side of the body, the fibres running obliquely downwards and backwards to end in an aponeurosis which passes dorsal to the **rectus abdominis** to be inserted into the linea alba.

iv. The **obliquus internus** lies beneath the **obliquus externus**, which must be removed in order to see it. It arises from the transverse processes of the vertebræ from the fourth backwards, and from the ilium. The fibres run downwards and forwards, and are inserted in front into the coracoid and sternum: some of the fibres surround, and are inserted into, the œsophagus and pericardium. The hinder two-thirds of the muscle pass dorsal to the tendon of the **obliquus externus** and are inserted, like it, into the linea alba.

2. **Muscles of the back.**

Pin out the frog on its belly: remove the skin, and clean the muscles in order.

i. The **depressor mandibulæ** is a broad triangular muscle which arises from the fascia covering the dorsal surface of the suprascapula: the fibres run downwards behind the tympanic membrane, and converge to be inserted into the angle of the lower jaw. The muscle by its contraction opens the mouth.

ii. The **cucullaris** is a small oblong muscle which, arising from the exoccipital near the middle line, runs backwards and outwards, and is inserted into the dorsal border of the suprascapula.

iii. The **latissimus dorsi** is a triangular muscle lying behind the **depressor mandibulæ**: it arises from the fascia dorsalis just behind the shoulder girdle, its origin being partly covered by the obliquus externus. The fibres run forwards and outwards, converging to be inserted by a long tendon into the deltoid ridge of the humerus.

Dissect away the **depressor mandibulæ** *and* **latissimus dorsi** *from their origins, and turn them down.*

iv. The **infraspinatus** arises from the dorsal surface of the suprascapula, partly overlapped by the **latissimus dorsi**: it runs outwards to be inserted into the deltoid ridge of the humerus: its action is to elevate the arm.

Lift up the suprascapula and note the muscles attaching it to the body, viz.,

v. The **retrahens** scapulæ, behind.

vi. The **levator anguli scapulæ**, in front.

Remove the suprascapula on one side, and clean the median longitudinal muscles of the back.

vii. The **extensor dorsi communis** is a longitudinal mass of muscle arising from the urostyle, and running forwards and slightly outwards: it is inserted into the ilium, into the transverse processes of the vertebræ, and in front into the posterior end of the skull. The anterior part is divided by transverse tendinous intersections.

viii. The **intertransversales** are small muscles, running between the transverse processes of the vertebræ, and lying beneath the **extensor communis**.

ix. The **glutæus** arises from the outer side of the posterior two-thirds of the ilium, and runs backwards to be inserted into the trochanter of the femur.

B. Muscles of the Head.

1. Muscles of the ventral surface of the head.

i. The **mylohyoid** or **submandibular** muscle is a flat sheet of muscle running across from one ramus of the mandible to the other, and divided down the middle line by a tendinous intersection. A narrow strip along the posterior border is commonly separated by a slight interval from the major or anterior part of the muscle.

Remove the **mylohyoid** *muscle, and the sternal portion of the* **pectoralis.**

ii. The **geniohyoid** is a narrow longitudinal band a short distance from the middle line: it arises from the lower jaw close to the chin, runs back on the ventral surface of the body of the hyoid, and divides posteriorly into two portions which are inserted respectively into the bony and the cartilaginous posterior processes of the hyoid.

iii. The **sternohyoid** is practically the anterior continuation of the **rectus abdominis.** It arises from the dorsal surface of the coracoid and clavicle, and is inserted into the ventral surface of the body of the hyoid, the tendon passing between the two divisions of the **geniohyoid.**

iv. The **hyoglossus** arises on either side from the posterior bony horn of the hyoid: the two muscles converge and meet each other in front of the larynx. In front of the larynx the muscle runs forward in the middle line as a stout band nearly to the chin: it then enters the tongue, and runs along it backwards to the tip.

v. The **petrohyoid** muscles are a set of five muscular bands which arise close together from the outer surface of the auditory capsule, and diverging in a fan-like manner, pass round the floor of the pharynx and œsophagus to be inserted in front into the median ventral line of the pharynx, and behind into the side of the hyoid. The first or most anterior band is a wide thin sheet of muscular tissue, while the four posterior portions are very narrow slips.

2. **Muscles of the side of the head.**

Remove the skin carefully from the side of the head and jaws, noticing how much more closely it is attached to the underlying parts than was the case in the body.

a. **Depressors of the lower jaw**; opening the mouth.

i. The **depressor mandibulæ** has been already seen and dissected. (See p. 57.)

b. **Elevators of the lower jaw**; shutting the mouth. These lie in the space between the auditory capsule and the eye.

ii. The **temporalis** arises from the upper surface of the auditory capsule, and passes outwards and downwards between the pterygoid and maxillary bones, and in front of the cartilaginous ring supporting the tympanic membrane, from which some of its fibres arise: it is inserted into the coronoid process of the lower jaw.

iii. The **pterygoideus** is a slender muscle placed just in front of the **temporalis** and partly covered by it: it arises from the side wall of the skull, and is inserted into the mandible further back than the temporalis, and very close to the joint.

iv. The **masseter** is a small muscle placed behind the **temporalis** : it arises from the quadratojugal and runs downwards and slightly backwards to be inserted into the outer surface of the mandible, just in front of the joint.

To see the insertions of these last three muscles the mouth should be opened widely.

3. **Muscles of the eyeball.**

Remove the temporal and pterygoid muscles carefully, dissecting them away from their origins, and then turning the muscles down and cutting them short close to their insertions. Remove also the lower jaw; pin the frog out on its back and dissect away carefully the mucous membrane of the roof of the mouth.

i. The **levator bulbi** is a thin sheet of muscle lying between the mucous membrane and the eye. Its fibres arise from the side of the skull, run outwards underneath the eye, and are inserted into the upper jaw. The muscle by its contraction serves to lift up the eyeball and so make it more prominent. Some of its fibres are inserted into the lower eyelid, which they serve to depress, acting as a **depressor palpebræ inferioris.**

Remove the **levator bulbi** *and clean the remaining muscles, dissecting them partly from the dorsal and partly from the ventral surface.*

a. The **recti muscles** are a group of four small muscles which arise close together from the inner and posterior angle of the orbit close to the optic foramen, and run forwards and outwards, diverging from one another, to be inserted into the bulb of the eye.

i. The **rectus superior** is inserted into the dorsal surface of the eyeball: it is seen best from above.

ii. The **rectus externus**, the most posterior of the four, is inserted into the posterior surface of the eyeball: it is seen best from the side or from below.

iii. The **rectus internus**, the longest of the four, runs forwards between the skull-wall and the eyeball, and is inserted into the inner or median surface of the eyeball: it is seen best from below.

iv. The **rectus inferior** is inserted into the under surface of the eyeball: it is seen best from below.

b. The **obliqui muscles** are two small muscles which arise close together from the palatine bone at the anterior end of the orbit, and run backwards to be inserted into the eyeball.

i. The **obliquus superior** is inserted into the dorsal surface of the eyeball just in front of the **rectus superior**: it is seen best from above.

ii. The **obliquus inferior** passes backwards beneath the **rectus internus**, and is inserted into the eyeball between it and the **rectus inferior**: it is seen best from below.

c. The **retractor bulbi**, or choanoid muscle, is a funnel-shaped muscle which lies within the four **recti** and embraces the optic nerve: it arises from the para-sphenoid, and is inserted into the eyeball. It is best exposed from below by carefully removing the **recti** muscles.

C. Muscles of the Hind-limb.

If the frog's leg be stretched back parallel to the longitudinal axis of the body, as in the act of swimming, we distinguish in it ventral and dorsal surfaces, an outer border in which is the projection of the knee, and an inner border along which is the bend of the knee. The outer border, which corresponds to the front of the leg in man, is called the **extensor surface**, inasmuch as the muscles which extend or straighten the leg lie along this edge: the inner border is the **flexor surface**. The "ventral" and "dorsal" surfaces only appear to be such in consequence of the extreme obliquity of the pelvic girdle: they are really anterior and posterior, and are better called **preaxial** and **postaxial**: they correspond respectively to the inner and outer surfaces of the human leg. If the foot be examined carefully, it will be seen that the first digit or "big toe" is on the preaxial side, and hence may be called the **preaxial digit**: while the fifth or "little toe" is on the postaxial side, and is therefore the **postaxial digit**.

1. Muscles of the thigh.

Remove the skin from one of the legs of the frog, and clean the muscles first of the preaxial and then of the postaxial surfaces.

a. Superficial muscles of the preaxial (apparent ventral) surface of the thigh.

i. The **sartorius** is a long narrow muscular band which crosses the thigh somewhat obliquely from the outer to the inner side. It arises from the iliac symphysis below the acetabulum, and is inserted into the inner side of the head of the tibia.

ii. The **adductor magnus** is a large muscle lying along the inner border of the **sartorius**, but passing beneath it at its distal end. It arises from the pubic and ischial symphyses, and passes under the **sartorius** to be inserted into the distal third of the femur.

iii. The **adductor longus** is a long narrow muscle lying along the outer side of the **adductor magnus**, and often completely hidden by the **sartorius**: it arises from the iliac symphysis beneath the **sartorius**, and unites a little way beyond the middle of the thigh with the **adductor magnus**.

iv. The **rectus internus major** is a large muscle lying along the inner side of the **adductor magnus** and of the **sartorius**. It arises from the ischial symphysis and is inserted into the head of the tibia.

v. The **rectus internus minor** is a narrow flat band of muscle running along the inner, or flexor margin of the thigh: it arises from a tendinous expansion connected with the ischial symphysis, and is inserted into the inner side of the tibia, just below its head.

b. Superficial muscles of the extensor surface of the thigh.

i. The **triceps extensor femoris**, the great extensor muscle of the thigh, arises by three distinct origins, which will be described separately, and is inserted into the tibia just below its head.

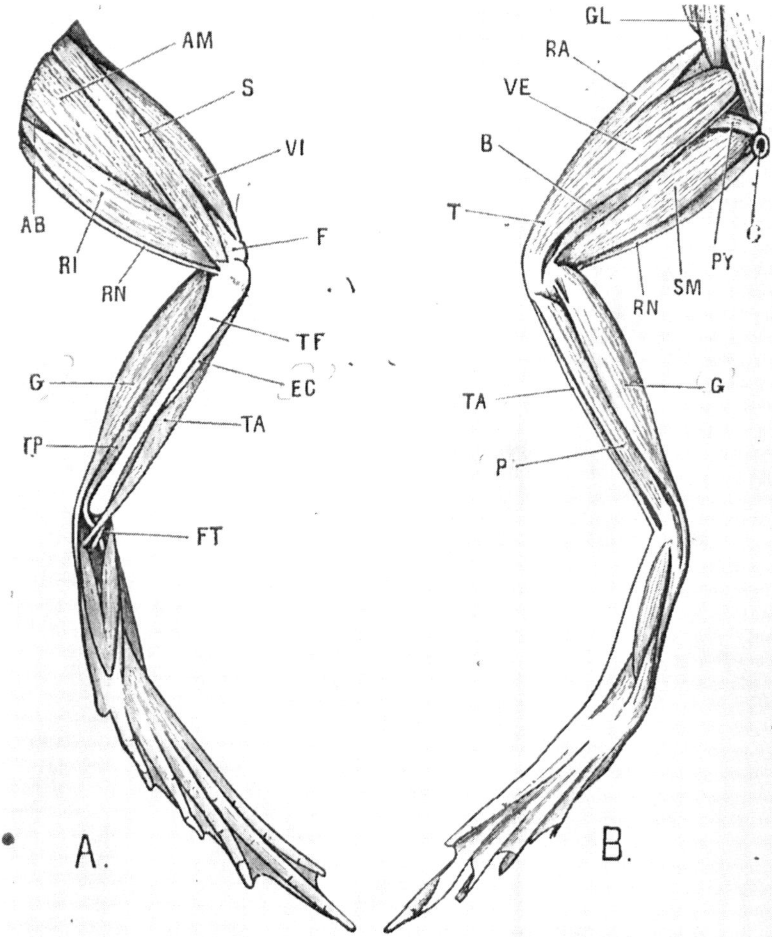

FIG. 11.—The superficial muscles of the frog's left hind-limb. A : from the preaxial surface ; B : from the postaxial surface.

AB, adductor brevis ; AM, adductor magnus ; B, biceps ; C, cloacal aperture ; EC, extensor cruris ; F, distal end of femur ; FT, tendon of flexor tarsi ; G, gastrocnemius ; GL, glutæus ; P, peroneus ; PY, pyriformis ; RA, rectus anticus femoris ; RI, rectus internus major ; RN, rectus internus minor ; S, sartorius ; SM, semimembranosus ; T, triceps extensor femoris ; TA, tibialis anticus ; TF, tibio-fibula ; TP, tibialis posticus ; VE, vastus externus ; VI, vastus internus.

α. The **rectus anticus femoris** forms the middle division of the **triceps**: it arises from the ventral border of the posterior third of the ilium, in front of the acetabulum: about half-way down the thigh it joins the next division.

β. The **vastus internus**, the preaxial division of the **triceps**, is a large muscle arising from the ventral and anterior border of the acetabulum, and lying in the thigh between the **sartorius** and the **rectus anticus**.

γ. The **vastus externus**, the postaxial division of the **triceps**, arises from the posterior edge of the dorsal crest of the ilium, and joins the other two divisions of the **triceps** about the junction of the middle and distal thirds of the thigh.

c. **Superficial muscles of the postaxial (apparent dorsal) surface of the thigh.**

i. The **glutæus** has been already noticed: it lies in the thigh between the **rectus anticus** and the **vastus externus.**

ii. The **biceps** is a long slender muscle which arises from the crest of the ilium just above the acetabulum: it lies in the thigh along the inner border of the **vastus externus**, and is inserted by a flattened tendinous expansion into the distal end of the femur and the head of the tibia.

iii. The **semimembranosus** is a stout muscle lying along the inner side of the **biceps**, between it and the **rectus internus minor.** It arises from the dorsal angle of the ischial symphysis just beneath the cloacal opening, and is inserted into the back of the head of the tibia. It is divided about its middle by an oblique tendinous intersection.

iv. The **pyriformis** is a slender muscle which arises from the tip of the urostyle, passes backwards and outwards between the **biceps** and the **semimembranosus**, and is inserted into the femur at the junction of its proximal and middle thirds.

d. Deep muscles of the thigh.

Lay the frog on its back and dissect the thigh from the preaxial surface. Separate the **adductor magnus** *and the* **rectus internus major** *with blunt instruments so as to expose the following muscles:*

i. The **semitendinosus** is a long thin muscle which arises by two heads; an anterior one from the ischium close to the ventral angle of the ischial symphysis and the acetabulum; and a posterior one from the ischial symphysis. The anterior head passes through a slit in the **adductor magnus** and unites with the posterior head in the distal third of the thigh. The tendon of insertion is long and thin, and joins that of the **rectus internus minor** to be inserted into the tibia just below its head.

Divide the **adductor magnus** *and the* **sartorius** *in the middle and turn the cut ends backwards and forwards, so as to expose the following muscles:*

ii. The **adductor brevis** is a short wide muscle, lying beneath the upper end of the **adductor magnus.** It arises from the pubic and ischial symphyses, and is inserted into the preaxial surface of the proximal half of the femur.

iii. The **pectineus** is a rather smaller muscle, lying along the outer (extensor) side of the **adductor brevis.** It arises from the anterior half of the pubic symphysis in front of the **adductor brevis,** and is inserted like it into the proximal half of the femur.

iv. The **ilio-psoas** arises by a wide origin from the inner surface of the acetabular portion of the ilium: it turns round the anterior border of the ilium, and crosses in front of the hip-joint, where for a short part of its course it is superficial between the heads of the **vastus internus** and of the **rectus anticus femoris**: it then passes down the thigh beneath these muscles, and is inserted into the back of the proximal half of the femur.

v. The **quadratus femoris** is a small muscle on the back of the upper part of the thigh: it arises from the ilium above the acetabulum, and from the base of the iliac crest: it lies beneath the **pyriformis** and behind the **biceps**, and is inserted into the inner surface of the proximal third of the femur between the **pyriformis** and the **ilio-psoas.**

vi. The **obturator** is a deeply situated muscle which arises from the whole length of the ischial symphysis and the adjacent parts of the iliac and pubic symphyses, and is inserted into the head of the femur close to the **glutæus.**

2. Muscles of the leg.

As in the thigh, we distinguish extensor and flexor surfaces, corresponding to the front and back of the leg in man; and also preaxial and postaxial surfaces, corresponding to the inner and outer sides of the human leg.

Lay the frog on its belly and commence the dissection from the postaxial surface.

i. The **gastrocnemius** is the large muscle forming the calf of the leg: it has two heads of origin, of which the larger arises by a strong, flattened tendon from the flexor surface of the distal end of the femur; while the smaller head, which joins the main muscle about one-fourth of its length below the knee, arises from the edge of the tendon of the **triceps extensor femoris** where it covers the knee. The muscle is thickest in its upper third, and tapering posteriorly ends in the strong **tendo Achillis,** which passes under the ankle-joint, being much thickened as it does so, and ends in the strong plantar fascia of the foot.

ii. The **tibialis posticus** arises from the whole length of the flexor surface of the tibia: it ends in a tendon which passes round the inner malleolus, lying in a groove in the lower end of the tibia, and is inserted into the dorsal surface of the astragalus.

iii. The **tibialis anticus** lies on the extensor surface of the leg: it arises by a long thin tendon from the lower end of the femur, and divides about the middle of the leg into two bellies which are inserted into the proximal ends of the astragalus and calcaneum respectively.

iv. The **extensor cruris** lies along the preaxial side of the **tibialis anticus**, partly covered by this and partly by the strong fascia of the leg. It arises by a long tendon from the preaxial condyle of the femur, runs in a groove in the upper end of the tibia, and is inserted into the extensor surface of the tibia along nearly its whole length.

v. The **peroneus** is a stout muscle which lies along the postaxial surface of the leg, between the **tibialis anticus** and the **gastrocnemius.** It arises from the distal end of the femur, and is inserted into the outer malleolus of the tibia and the proximal end of the calcaneum.

CHAPTER V.

THE NERVOUS SYSTEM OF THE FROG.

THE nervous system consists of the following parts:

1. A central portion, the **brain** and **spinal cord**, which lies in the cartilaginous and bony tube formed by the cranium and vertebral column, and which is the centre where sensations are felt, and whence motor impulses causing the muscles to contract take their origin.
2. A peripheral portion, the **nerves** themselves, which connect the central portion with the skin, sense organs, muscles, viscera, etc., and serve to convey sensory impulses from these parts to the brain and cord, or motor impulses from the central organs to the muscles. These two functions are fulfilled by different nerves, which may accordingly be distinguished as (*a*) **afferent** or **sensory nerves**, conveying impulses *to* the brain or cord and (*b*) **efferent** or **motor nerves**, conveying impulses *from* the brain or cord.

FIG. 12.—The nervous system of the edible frog (*Rana esculenta*), from the ventral surface. (From Ecker.)

F, facial nerve; G, ganglion of pneumogastric nerve; He, cerebral hemisphere; Lc, optic tract; Lop, optic lobe; M, boundary between medulla oblongata and spinal cord; M 1–10, the spinal nerves; MS, connection between fourth spinal nerve and sympathetic chain; N, nasal sac; Ni, sciatic nerve; No, crural nerve; o, eyeball; S, trunk of sympathetic; S 1–10, the sympathetic ganglia; Sp, continuation of sympathetic into head.

I, olfactory nerve; II, optic nerve; III, motor oculi; IV, fourth nerve; V, trigeminal and facial nerves; Va, ophthalmic branch of trigeminal; Vc, maxillary branch of trigeminal; Vd, mandibular branch of trigeminal; Ve, hyomandibular branch of facial; Vg, Gasserian ganglion; Vs, upper end of sympathetic trunk, in connection with Gasserian ganglion; VI, abducens nerve; VII, facial nerve; VIII, auditory nerve; X, glossopharyngeal and pneumogastric nerves; X1, ramus anterior of glossopharyngeal; X2, ramus posterior of glossopharyngeal; X 3–4, branches of pneumogastric.

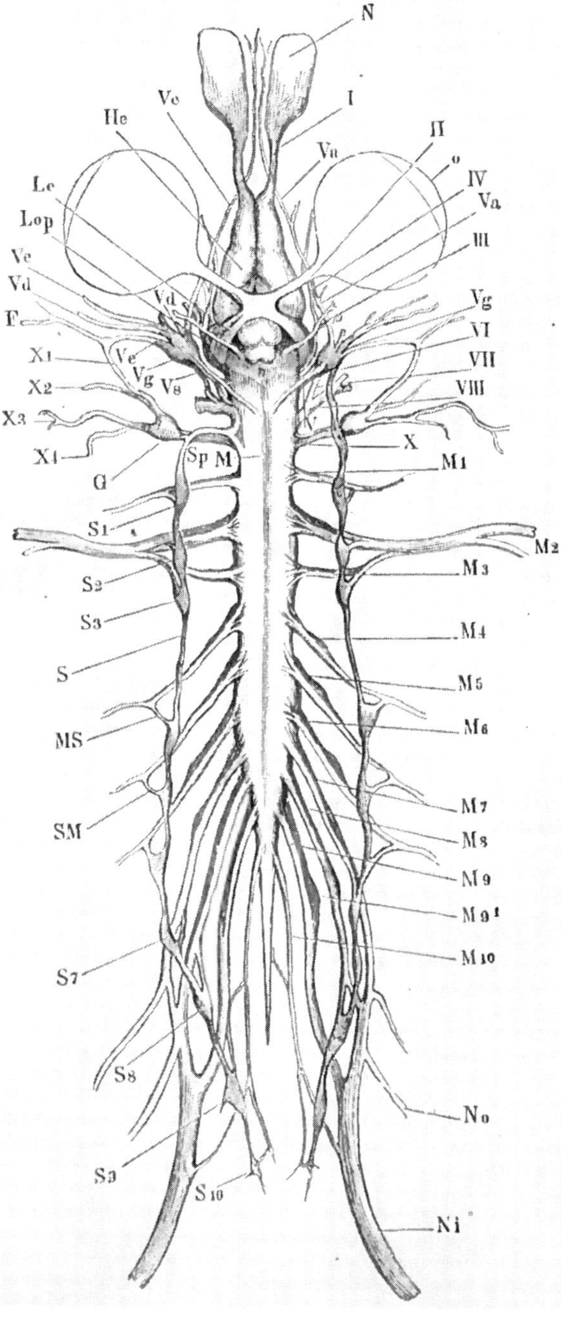
N
Vc
I
He
Va
II
o
IV
Le
Va
Lop
III
Ve
Vd
Vd
Vg
F
VI
X1
Ve
Vg
V8
VII
X2
VIII
X3
X
X4
Sp M
M1
G
S1
M2
M3
S2
S3
M4
S
M5
M6
MS
M7
SM
M8
M9
M9¹
M10
S7
S8
No
S9
S10
N1

A special set of nerves in connection with the bloodvessels and viscera forms the **sympathetic nervous system.**

For the dissection of the nervous system specimens should be taken which have been in strong spirit for two or three days, and in which the brain has been exposed to the action of the spirit by removal of the roof of the skull.

A. The Central Nervous System.

This is divisible into an anterior portion—the **brain**—lying in the cavity of the cranium; and a posterior portion—the **spinal cord**—which lies in the neural canal of the vertebral column. There is no sharp line of demarcation between the two portions, which are directly continuous with each other.

If the brain and spinal cord have not already been exposed, clear away the dorsal muscles from both sides of the spine: cut through the occipito-atlantal membrane, flexing the frog's head slightly to make the membrane tense, and being careful not to injure the brain beneath it. Introduce one blade of the scissors into the cranial cavity, with the flat surface of the blade parallel to the back of the frog, and keeping as close to the roof of the skull as possible. Cut carefully through the side walls of the skull, first on one side and then on the other. Turn the roof of the skull forwards with forceps, and remove it altogether.

Similarly cut through and remove the neural arches of the vertebræ one by one, from before backwards.

Examine and draw the central nervous system in situ, *showing its several parts.*

I. The Brain.

a. **The dorsal surface of the brain**: *note from before backwards the following parts, removing the pigmented membrane* (**pia mater**) *covering the several parts as you come to them.*

i. The **olfactory lobes**, which form the most anterior portion of the brain, are united together in the median plane: they give off the olfactory nerves from their anterior ends, and are separated behind by slight constrictions from the hemispheres.

ii. The **cerebral hemispheres** are a pair of smooth ovoid bodies which touch each other in the median plane but are not fused together.

iii. The **thalamencephalon** is a lozenge-shaped portion lying immediately behind the hemispheres and between their diverging posterior ends: it is covered by a thick vascular membrane—the **choroid plexus**—over which passes the stalk of the **pineal body**, a small body adherent to and generally removed with the roof of the skull. On removing the **choroid plexus** a slit-like hole is

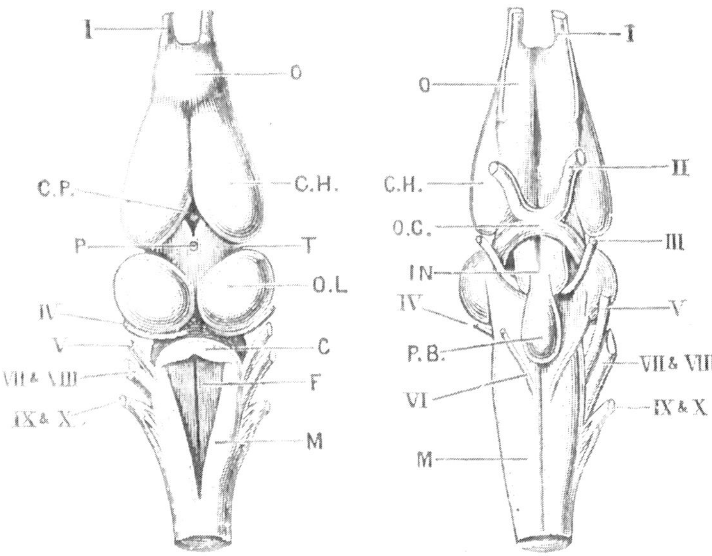

FIG. 13.—The brain of the frog: dorsal surface. × 4.
FIG. 14.—The brain of the frog: ventral surface. × 4.
C, cerebellum; CH, cerebral hemisphere; CP, choroid plexus of third ventricle; F, fourth ventricle; IN, tuber cinereum; M, medulla oblongata; O, olfactory lobe; OC, optic chiasma; OL, optic lobe; P, stalk of pineal body; PB, pituitary body; T, thalamencephalon.
I, olfactory nerve; II, optic nerve; III, third or motor oculi nerve; IV, fourth nerve; V, fifth or trigeminal nerve; VI, sixth nerve; VII and VIII, combined root of facial and auditory nerves; IX and X, combined root of glossopharyngeal and pneumogastric nerves.

left in the roof of the thalamencephalon. The vessels of the plexus, covered by a thin layer of nervous matter, hang into the third ventricle, or cavity of the thalamencephalon. The thickened sides of the thalamencephalon are the **optic thalami**.

iv. The **optic lobes** are a pair of prominent ovoid bodies touching each other in the median plane and forming the widest part of the brain: the pia mater covering them is very strongly pigmented.

v. The **cerebellum** is a narrow transverse band immediately behind the optic lobes.

vi. The **medulla oblongata** is the part of the brain behind the cerebellum: it is widest in front and gradually tapers towards its posterior end, where it is continuous with the spinal cord. It is covered by a triangular and very vascular membrane, beneath which lies the fourth ventricle.

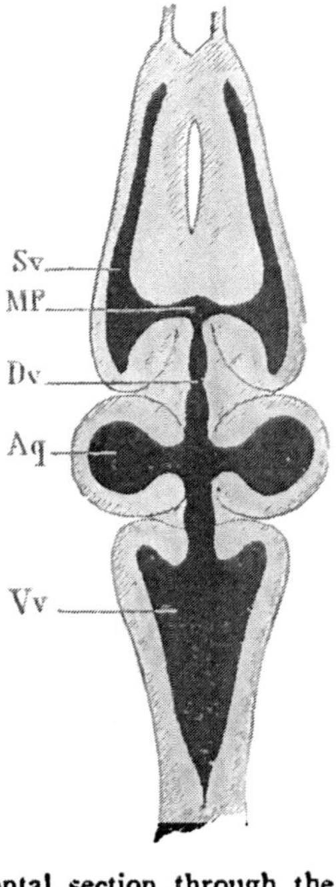

FIG. 15.—A horizontal section through the brain of the frog, to show the internal cavities. (From Ecker.)

Aq, ventricles of the optic lobes; Dv, third ventricle; MF, foramen of Monro; Sv, lateral ventricle; Vv, fourth ventricle.

b. The cavities of the brain.

Slice off the upper surface of the brain horizontally so as to expose the several cavities or ventricles, without removing it from the skull. These cavities are merely parts of, or outgrowths of, the original central canal of the neural tube of the embryo. (*Cf.* Chap. VIII.)

i. The **lateral ventricles** extend through the whole length of the cerebral hemispheres and a short way into the olfactory lobes.

ii. The **third ventricle** is situated in the thalamencephalon: it opens in front through the **foramina of Monro** into the lateral ventricles: the stalk of the pineal body opens into it above; and in the hinder part of its floor is a conical depression, the **infundibulum**.

iii. The **aquæductus Sylvii** or **iter a tertio ad quartum ventriculum** is a narrow passage leading from the third to the fourth ventricle: it communicates above with the cavities or ventricles of the optic lobes, which are hollow.

iv. The **fourth ventricle** is the large triangular cavity in the medulla, already exposed by removal of the vascular membrane covering it.

c. The ventral surface of the brain.

Cut through the medulla at the level of the hinder end of the skull: carefully remove the brain from the cranial cavity, noting the several nerves arising from it, and cutting through these as far from the brain as possible. Lay the brain on its dorsal surface, examine and draw the ventral surface, showing the following parts:

i. The **optic chiasma** is formed by the decussation of the roots of the optic nerves; the point of crossing being opposite the hinder ends of the hemispheres, and immediately in front of the infundibulum.

Trace back the optic nerves behind their point of crossing to their origins from the optic lobes.

ii. The **tuber cinereum** is a small median swelling immediately behind the optic chiasma, caused by the depression of the floor of the third ventricle to form the infundibulum. It is divided by a median groove into right and left halves.

iii. The **pituitary body** is a flattened ovoid sac, lying behind, and continuous with, the tuber cinereum. It is almost certain to be left behind in the skull on removing the brain, in which case the infundibulum will be seen torn across.

iv. The **crura cerebri** are two dense white columns of nervous matter, lying at the base of the optic lobes, and partly hidden by the pituitary body: they serve to connect the hemispheres with the medulla and spinal cord.

v. The **ventral fissure of the brain** is a median longitudinal groove on the ventral surface of the hinder part of the brain: it is continuous with a similar groove on the ventral surface of the spinal cord.

II. The Spinal Cord.

The spinal cord is a somewhat flattened band, presenting brachial and lumbar enlargements opposite the points of origin of the nerves for the fore and hind limbs respectively, and slightly constricted between these two points. About the level of the sixth or seventh vertebra the cord narrows rapidly to form a fine thread, the **filum terminale**, which extends back into the canal of the urostyle.

The tubular character of the spinal cord is best seen on making transverse sections of it. (See p. 85.)

B. The Peripheral Nervous System.

1. The Spinal Nerves. Ten pairs of nerves arise from the sides of the spinal cord; each nerve arising by two roots, a ventral or "anterior," and a dorsal or "posterior," which unite at their point of exit from the vertebral canal through the intervertebral foramen: just before their union the dorsal root bears a ganglionic swelling.

Within the vertebral canal the roots of the anterior spinal nerves run nearly transversely outwards, so as to leave the canal opposite their points of origin from the spinal cord. The roots of the middle and posterior nerves, owing to the vertebral column being of greater length than the part of the cord belonging to it, pass obliquely backwards to their points of exit: and in the case of the hindmost nerves, the roots run backwards within the vertebral canal some distance before reaching their foramina of exit: the bundle formed by these roots, together with the filum terminale, is spoken of as the **cauda equina.**

a. The spinal nerves outside the vertebral canal.

Lay the frog on its back: cut through and pin out the body-walls, and remove the abdominal viscera. Note the spinal nerves, seen as white cords at the sides of the vertebral column. Clean the nerves on one side and follow them to their distribution.

Each nerve divides, directly after the union of its two roots, into a small dorsal branch, and a much larger ventral branch.

i. The ventral branches.

1. The **hypoglossal,** or first spinal nerve, leaves the vertebral canal between the first and second vertebræ, and then runs forwards on the under surface of the head beneath the mylohyoid and in the substance of the geniohyoid muscle to the chin, where it enters the tongue, in which it ends. It supplies the muscles of the tongue and floor of the mouth, and also some of the muscles of the back and shoulder. (Fig. 16.) It frequently sends one or two branches to the brachial plexus.

2 and 3. The second and third spinal nerves leave the canal between the second and third, and third and fourth vertebræ respectively. The **brachial plexus** is formed by the fusion of (*a*) one or two branches from the hypoglossal nerve, (*b*) the second spinal nerve, (*c*) a branch from the third spinal nerve. These unite together to form the **brachial nerve,** which gives off a large **coracoclavicular** branch to the shoulder muscles and then runs down the arm, supplying it with muscular and cutaneous branches, and divides just above the elbow into the **radial** and **ulnar nerves,** supplying the forearm and hand.

4, 5, and 6. The fourth, fifth, and sixth spinal nerves are small and supply the muscles and skin of the body-wall. They leave the vertebral canal between the fourth and fifth, fifth and sixth, and sixth and seventh vertebræ respectively.

7, 8, and **9.** The seventh, eighth, and ninth spinal nerves together form the **sciatic plexus** The roots of these three nerves within the vertebral canal form the main part of the **cauda equina.** The seventh nerve leaves the canal between the seventh and eighth vertebræ, the eighth nerve between the eighth and ninth vertebræ, and the ninth nerve between the ninth or sacral vertebra and the urostyle. Outside the vertebral canal the three nerves unite together opposite the middle of the urostyle to form the sciatic plexus, from which branches are given to the large intestine, bladder, oviducts, etc. Just before joining the plexus the seventh nerve gives off the **ileohypogastric** and **crural nerves,** supplying the muscles and skin of the abdomen and thigh. Beyond the plexus is the large **sciatic nerve,** which runs down the thigh, giving branches to it, and dividing a short distance above the knee into the **tibial** and **peroneal nerves** supplying the leg and foot.

The relative sizes of the nerves forming the sciatic plexus, and the mode of their union with one another are subject to considerable individual variation in different frogs.

10. The **coccygeal** or tenth spinal nerve emerges through a small hole in the side of the urostyle near its anterior end. It gives branches to the bladder, cloaca, and other adjacent parts, and contributes a branch to the sciatic nerve.

ii. The dorsal branches of spinal nerves 2 to 8 inclusive are to be found traversing the dorsal lymph-sac in their passage to the skin.

b. The spinal nerves within the vertebral canal.

Cut away with scissors the centra of the vertebræ one by one, to expose the spinal cord and the roots of the spinal nerves from the ventral surface. Note the following points :

i. The roots of the nerves: dorsal and ventral.
ii. The obliquity of the middle and posterior roots.
iii. The **cauda equina**, formed by the roots of the hinder nerves together with the **filum terminale**.
iv. The **ganglia** on the dorsal roots as they pass through the intervertebral foramina: these are covered on their ventral surfaces by whitish calcareous patches (the **periganglionic glands** or "glands of Swammerdam") which form conspicuous objects on either side of the vertebral column. *Remove these patches carefully to see the ganglia.*

II. The Sympathetic Nervous System. This consists of a longitudinal nervous band on each side of the body, connected by branches with the several spinal nerves. The two main sympathetic trunks lie, in front, close to the dorsal surface and alongside the vertebral column: further back they are in close relation with the dorsal aorta, alongside which they run.

Each sympathetic trunk receives a branch from each of the spinal nerves of its side, and at the junction of each of these branches with the main trunk there is a ganglionic enlargement.

The coccygeal or tenth spinal nerve, unlike the others, is connected with the sympathetic by more than one branch: the actual number of these branches is not constant, but is said to vary from two up to as many as twelve.

From the sympathetic ganglia nerves are given off to the bloodvessels and viscera, the chief ones being the following:

i. The **cardiac plexus** is formed by nerves arising from the first sympathetic ganglion: the plexus is a meshwork of nerves on the auricles, and around the great vessels at their openings into the heart.
ii. The **solar plexus** lies on the dorsal surface of the stomach: the nerves are derived mainly from the third, fourth, and fifth ganglia.

Hepatic, renal, genital, hæmorrhoidal, and vesical plexuses also exist in connection with the liver, kidney, reproductive organs, large intestine, and bladder respectively.

III. The Cranial Nerves. There are ten pairs of cranial nerves in the frog, which are numbered in order from before backwards. (See Figs. 12, 14 and 16.)

To dissect the cranial nerves expose the brain by removing the roof of the skull as already described, and then follow the special instructions given in the case of the more important nerves.

1. The **olfactory nerve,** the special nerve of smell, arises from the anterior end and outer side of the olfactory lobe, and is distributed to the membrane lining the nasal cavity.

To see the course and distribution of the olfactory nerve, dissect from the dorsal surface, removing the roof of the anterior part of the skull, including the nasal bone.

2. The **optic nerve,** the nerve of sight, arises from the side

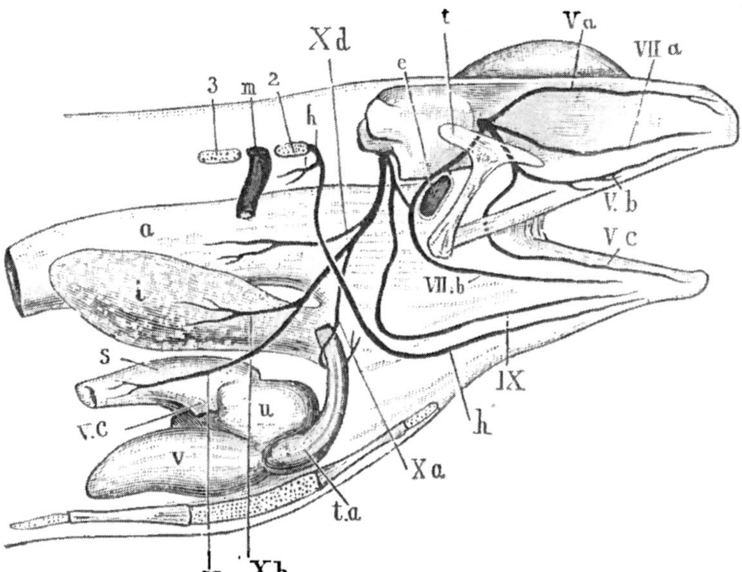

FIG. 16.—The trigeminal, facial, glossopharyngeal, and pneumogastric nerves of the frog, dissected from the right side.

a, stomach; *e*, Eustachian tube; *h*, hypoglossal nerve; *i*, lung; *m*, second spinal nerve; *s*, sinus venosus; *t*, squamosal; *t.a*, truncus arteriosus; *u*, right auricle; *v*, ventricle; *v.c*, right anterior vena cava; Va, ophthalmic branch of trigeminal; Vb, maxillary branch of trigeminal; Vc, mandibular branch of the trigeminal; VIIa, palatine branch of facial; VIIb, hyoidean branch of facial; IX, glossopharyngeal; Xa, laryngeal branch of pneumogastric; Xb, pulmonary branch of pneumogastric; Xc, cardiac branch of pneumogastric; Xd, gastric branch of pneumogastric; 2, transverse process of second vertebra; 3, transverse process of third vertebra.

of the brain just below the optic lobe, partially crosses over at the optic chiasma on the under surface of the brain, and then runs outwards to the eyeball.

The course of the optic nerve has been fully seen in previous dissections.

3. The **motor oculi** is a small nerve arising from the ventral surface of the brain close to the median line and between the crura cerebri. It supplies four of the muscles moving the eyeball; viz., the rectus superior, rectus internus, rectus inferior, and obliquus inferior.

Owing to its small size the third nerve is not easily made out in the frog.

4. The **fourth** or **pathetic nerve** is a very slender nerve, arising from the dorsal surface of the brain between the optic lobe and the cerebellum, and supplying the obliquus superior muscle of the eyeball.

The fourth nerve is too small to be dissected satisfactorily in the frog.

5. The **trigeminal** is the largest of the cranial nerves in the frog. It arises from the side of the anterior part of the medulla, and runs outwards and forwards to the skull wall: just before reaching this it expands into a large swelling—the **Gasserian ganglion.** It then passes through the skull wall immediately in front of the auditory capsule, and divides at once into two main branches.

    i. The **ramus ophthalmicus** runs forwards through the orbit, lying close to its inner side, between the skull wall and the eye. It passes beneath the rectus superior, but above all the other muscles of the eyeball and the optic nerve. At the anterior end of the orbit it divides into two branches, which pass through the walls of the nasal capsule, and supply the skin of the fore part of the head.

To trace this branch, dissect from the dorsal surface: cut away carefully with scissors the side wall of the cranium: cut through and turn aside the rectus superior, and find the nerve running close alongside the skull wall, between it and the eyeball. Trace it forwards to the nose.

ii. The **ramus maxillo-mandibularis** runs directly outwards behind the eyeball, in front of the auditory capsule and between the temporal and pterygoid muscles. After a very short course it divides into the maxillary and mandibular nerves.

To trace this nerve and its branches remove the squamosal bone carefully, and find the nerve lying on the pterygoid muscle and immediately behind the eye. Follow the nerve behind the pterygoid and temporal muscles to the skull, removing the muscles if necessary; and then trace the branches outwards to their distribution.

α. The **ramus maxillaris** runs forwards and outwards in the floor of the orbit, behind and below the eyeball, to the margin of the upper jaw, which it reaches about midway along its length: it then ends in branches which run along the jaw, some forwards and some backwards, supplying the upper lip, the lower eyelid, and other neighbouring parts.

β. The **ramus mandibularis** runs parallel to and behind the **ramus maxillaris** as far as the outer border of the eyeball, giving branches to the temporal and pterygoid muscles: it then turns backwards, outwards, and downwards, and passing across the inner side of the upper jaw, reaches the outer surface of the mandible just behind the insertion of the temporal muscle: it then runs forwards along the outer side of the lower jaw to the chin, supplying the lower lip and the muscles of the floor of the mouth.

6. The **abducens** is a very slender nerve which arises from the ventral surface of the medulla close to the median line, and a short way behind the pituitary body. It passes either through, or in very close contact with, the Gasserian ganglion, and entering the orbit supplies the retractor bulbi and the rectus externus muscles.

The abducens nerve is too small to be made out satisfactorily in the frog.

7. The **facial nerve** arises from the side of the medulla immediately behind the trigeminal nerve, and passes forwards to the skull wall, where it is very closely connected with the Gasserian ganglion. It passes through the skull wall immediately behind and in close company with the trigeminal nerve, and divides at once into its two main branches.

i. The **ramus palatinus** runs forwards in the floor of the orbit a short distance from the side wall of the skull, and immediately upon the mucous membrane of the roof of the mouth. Near the anterior end of the orbit it divides into two branches, one of which runs outwards and anastomoses with the ramus maxillaris of the trigeminal nerve, while the other runs forwards to the anterior part of the roof of the mouth. It supplies the mucous membrane of the roof of the mouth.

Dissect this nerve from the ventral surface. Cut away the lower jaw: carefully remove the mucous membrane of the roof of the mouth, and find the nerve lying on the ventral surface of the eyeball and its muscles, and running parallel to and a short distance from the skull wall. Trace it backwards and forwards.

ii. The **ramus hyomandibularis** runs outwards and backwards round the front end of the auditory capsule; it then crosses over the inner end of the columella and turns downwards in the posterior wall of the Eustachian tube to the angle of the mouth, giving branches to the tympanic membrane and to the articulation of the mandible. It then divides into two branches.

The exposure of the above nerve, which is not easy, may be performed thus: remove the shoulder girdle of one side; also the depressor mandibulæ and temporalis muscles: open the cranial cavity as before, to expose the brain: remove the tympanic membrane and clean the outer end of the columella. Cut away carefully the roof of the auditory capsule by a horizontal cut, just above the level of the columella: find the facial nerve running round the front end of the auditory capsule and in close contact with it, and trace it back over the columella and down to the angle of the mouth.

α. The **ramus mandibularis** runs forwards in the floor of the mouth, lying along the inner edge of the lower jaw and between the mylohyoid muscle and the skin, as far forward as the chin.

Dissect from the ventral surface: remove the skin from the under surface of the floor of the mouth, and find the nerve running along the inner border of the mandible.

β. The **ramus hyoideus** is the posterior and larger of the two branches: it runs forwards in the floor of thé mouth along the anterior cornu of the hyoid, supplying its muscles.

8. The **auditory nerve**, the nerve of hearing, arises from the side of the medulla immediately behind and in close contact with the root of the facial nerve: it enters the auditory capsule and ends in the internal ear.

9. The **glossopharyngeal nerve** arises from the side of the medulla behind the auditory nerve, by a root common to it and the tenth nerve: it leaves the skull by an aperture immediately behind the auditory capsule, and divides behind the capsule into two branches.

i. The **ramus anterior** runs downwards and forwards round the hinder border of the auditory capsule and beneath the depressor mandibulæ muscle to join the facial nerve just after it has crossed over the columella.

The dissection already made for the ramus hyomandibularis of the facial nerve will show also the above branch of the glossopharyngeal.

ii. The **ramus posterior** runs downwards and forwards to the ventral wall of the pharynx, passing beneath the fourth division of the petrohyoid muscle but superficial to the others; it runs just behind and parallel to the anterior cornu of the hyoid. On reaching the floor of the mouth it crosses the hypoglossal nerve obliquely, lying dorsal to it and then runs forwards in a peculiarly sinuous course, close to the middle line and between the geniohyoid and hyoglossus muscles, to the base of the tongue, which it enters and in which it ends. It supplies a petrohyoid muscle, and the mucous membrane of the pharynx and tongue.

The exposure of the first part of the above nerve is best performed from the side, and is much facilitated by distending the œsophagus and pharynx with a cork or roll of paper. Its course along the floor of the mouth to the tongue should be dissected from the ventral surface.

10. The **pneumogastric** or **vagus nerve** arises, as already noticed, in common with the glossopharyngeal. It leaves the skull by the same aperture as the ninth nerve, and immediately outside the skull presents a ganglionic enlargement: it gives off dorsal branches to the muscles of the back, and then runs backwards and downwards round the side wall of the pharynx, running along the hinder border of the fourth division of the petrohyoid muscle; behind this muscle it divides into its main branches, which are as follows:

   i. The **ramus laryngeus** or **recurrens** loops round the posterior cornu of the hyoid and round the pulmocutaneous artery close to its origin from the aortic trunk: it then passes inwards, dorsal to the artery, to the middle line, where it ends in the larynx.

   ii. The **ramus cardiacus** passes dorsal of the pulmonary artery to the interauricular septum of the heart, and to the sinus venosus.

   iii. The **rami pulmonales** follow the course of the pulmonary artery to the lung, in which they end.

   iv. The **rami gastrici**, usually two in number, run through the partial diaphragm formed by the anterior fibres of the obliquus internus muscle, and end in the walls of the stomach.

The dorsal portions of the several branches of the pneumogastric nerve are best exposed from the side; to see them properly, the shoulder girdle and fore limb must be removed and the œsophagus well distended: the terminal branches must be dissected from the ventral surface.

IV. The Cranial Portion of the Sympathetic Nervous System.

The main sympathetic trunk of each side extends forwards in front of the first ganglion, and enters the skull at the foramen in the exoccipital bone through which the glossopharyngeal and pneumogastric nerves pass out: it is connected with the pneumogastric nerve, and then runs forwards within the skull to the Gasserian ganglion of the trigeminal nerve, in which it ends.

C. Histology of Nerves.

Nervous matter consists histologically of elements of two kinds, **nerve cells** and **nerve fibres**. The nerve cells are branching nucleated cells, connected by their processes with one another and with the nerve fibres. The nerve cells are the centres whence impulses originate, while the nerve fibres serve to convey these impulses from one part to another. A local accumulation of nerve cells is called a ganglion.

1. **Nerve Fibres** are of two kinds, medullated and non-medullated.

   a. **Medullated nerve fibres** form almost the whole of the cranial and spinal nerves, and a large part of the brain and spinal cord.

Take a small piece of the sciatic or some other nerve from a freshly killed frog: spread it out and tease it in a drop of normal salt solution: examine with low and high powers: note

i. The **nerve fibres**: unbranched.

ii. The **perineurium**, or connective tissue binding the nerve fibres into bundles, or "nerves."

In each nerve fibre note

iii. The **primitive sheath**, or **sheath of Schwann**: a very delicate external investment, seen with difficulty, and only in certain places.

iv. The **medullary sheath**: a thick fatty layer within the primitive sheath; it is highly refractive, and gives the nerve fibre its double contour.

Tease in glycerine a small piece of nerve that has been treated with osmic acid: examine with the high power a single nerve fibre: note the following points:

i. The **medullary sheath** is stained darkly in consequence of its fatty nature: it is interrupted at intervals by the nodes of Ranvier.

ii. The **nodes of Ranvier** are spots where the medullary sheath is absent, and the primitive sheath forms constrictions touching the axis cylinder.

iii. The **axis cylinder** is the central cylindrical rod, the essential part of the nerve fibre: it is clearly visible at the nodes, and is much less deeply stained than the medullary sheath.

iv. **Nuclei** are seen projecting into the medullary sheath about midway between the nodes.

Tease a small piece of fresh nerve in chloroform: this will partially dissolve the fatty medullary sheath and so render the primitive sheath and axis cylinder more clearly visible: note

i. The **primitive sheath,** or **sheath of Schwann.**

ii. The **axis cylinder.**

b. **Non-medullated nerve fibres** occur chiefly in the sympathetic nerves: they branch and anastomose; and they have no medullary sheath.

2. **Nerve cells**: *Tease in glycerine a small fragment of the ventral cornu of the spinal cord of the ox (lumbar region): cover, and examine with low and high powers: note*

i. The **nerve cells**: large nucleated cells with many branching processes.

ii. The **nerve fibres.**

iii. The fine connective tissue binding the several parts together.

3. **Structure of the Spinal Cord.**

Take one of the prepared transverse sections of spinal cord of frog; mount in balsam, and examine with low and high powers: note the following points:

a. **With the low power.**

i. The section is bilaterally symmetrical, and oval in shape; the transverse diameter considerably exceeding the vertical.

ii. The **ventral** or "**anterior**" **fissure** is a broad and shallow median cleft: there is in the frog no distinct dorsal fissure.

iii. The **white matter** forms the outer part of the cord, and is chiefly composed of medullated nerve fibres.

iv. The **grey matter** forms the central part of the cord, and is composed of a dense network of non-medullated nerve fibres, in which are imbedded numerous nerve cells.

v. The **cornua** are the processes, ventral or "anterior," and dorsal or "posterior," into which the grey matter is produced on either side.

vi. The **central canal** of the cord lies in the median plane, nearer the ventral than the dorsal surface.

vii. The **nerve roots** are only seen if the section happens to pass through their points of origin.

α. The **dorsal** or "**posterior**" **root** is connected with the dorsal cornu of the grey matter; it is a single thick band of nerve fibres.

β. The **ventral** or "**anterior**" **root** is connected with the ventral cornu of the grey matter; it consists of a number of very slender bands of nerve fibres.

b. With the high power.

i. The **ganglion cells** are large nucleated branched cells lying in groups in the grey matter: they are largest and most numerous in the ventral cornua.

ii. The **neuroglia** is a delicate network of connective tissue fibres and cells, penetrating and supporting all parts of the cord, and continuous at the surface with the pia mater.

iii. The **pia mater** is the delicate connective tissue membrane closely ensheathing the cord.

iv. The **bloodvessels** of the cord are small and numerous: they enter from the pia mater.

v. The central canal is lined by a **ciliated epithelium.**

CHAPTER VI.

THE EYE AND EAR.

A. The Eye of the Frog.

1. *Remove the eye from a freshly killed frog: snip off with scissors the muscles of the eyeball: note the following points:*

- i. The **shape**. The eyeball is flattened on its outer side, more convex on the inner or deeper side.
- ii. The **sclerotic** is the firm outer wall of the eyeball, formed of hyaline cartilage, and dense white connective tissue.
- iii. The **cornea** is the transparent patch on the outer side of the eye through which the light enters: it is continuous at its margin with the sclerotic.
- iv. The **iris** is a pigmented ring placed behind the cornea and seen through it: it acts as a diaphragm, limiting the amount of light that enters the eye.
- v. The **pupil** is the aperture surrounded by the iris, which serves to admit the light to the interior of the eye.
- vi. The **optic nerve** is seen piercing the sclerotic to enter the eyeball on its inner side.

2. *Place the eye under water and divide it with scissors into two halves by a cut passing through the middle of the cornea and through the sclerotic close to the optic nerve, so as to lay open completely the interior of the eye: note the following points:*

- i. The **lens** is a firm, solid, transparent body, just behind the iris and attached to its outer margin: it is more convex on its inner than on its outer surface.

ii. The **anterior chamber of the eye** is the space between the cornea and the lens: it is small and contains the **aqueous humour,** a watery fluid.

iii. The **posterior chamber of the eye** is a large space behind the lens: it is filled by the **vitreous humour,** a gelatinous body.

iv. The **choroid** is the black pigmented layer lining the sclerotic, and continuous in front with the iris.

v. The **retina** is a delicate transparent membrane lining the eyeball. It is readily detached from the choroid, except at the point where the optic nerve enters.

B. The Eye of the Sheep or Ox.

1. *Dissect off the muscles of the eyeball, and the fat which surrounds the optic nerve: note the following points:*

i. The **shape.** The eyeball is more spherical than in the frog.

ii. The **sclerotic** covers about five-sixths of the eyeball: it is tough, white, and opaque.

iii. The **cornea,** which covers the outer sixth of the eyeball, is circular, transparent, and continuous at its margin with the sclerotic: it is more convex than the sclerotic.

iv. The **conjunctiva** is a delicate epithelial layer, continuous with the skin, covering the front of the cornea and part of the sclerotic.

v. The **iris** is the oval pigmented ring seen through the cornea.

vi. The **pupil** is the central, oval, or dumb-bell shaped aperture surrounded by the iris.

vii. The **optic nerve** is a thick white bundle of nerve fibres piercing the sclerotic at the back of the eye.

2. *Cut all round the cornea, with stout scissors, about $\frac{1}{8}$ inch*

from its junction with the sclerotic: remove the cornea: take care not to squeeze the eye, or the lens will be driven out instantly: note:

i. The **aqueous humour**: the transparent watery fluid filling the **anterior chamber of the eye** and escaping when the cornea is removed.

ii. The **lens**.

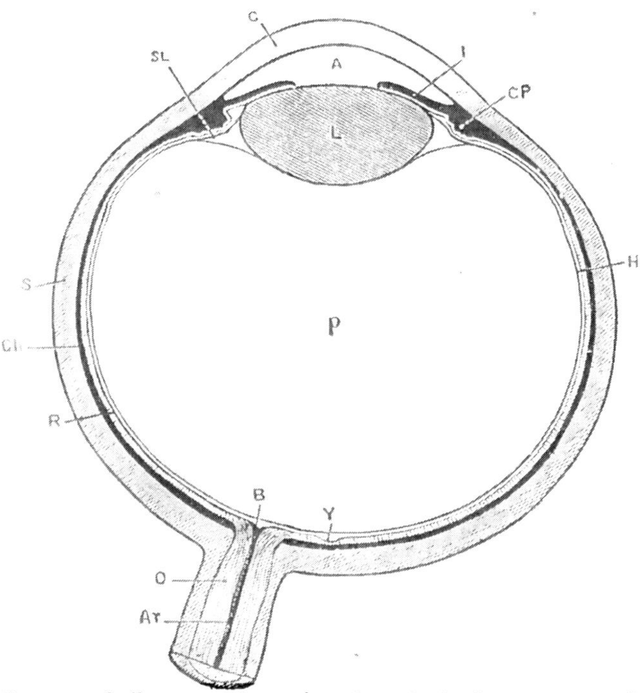

FIG. 17.—A diagrammatic section through the human eye passing through the centres of the cornea and lens, and through the yellow spot and point of entrance of the optic nerve.

A, anterior chamber; Ar, central artery of retina; B, blind spot; C, cornea; Ch, choroid; CP, ciliary processes; H, hyaloid membrane, enclosing the vitreous humour; I, iris; L, lens; O, optic nerve; P, posterior chamber; R, retina; S, sclerotic; SL, suspensory ligament; Y, yellow spot. (The yellow spot is absent from a beast's eye.)

3. *Pass the handle of a scalpel under the cut edge of the cornea, between it and the iris, and carefully separate the sclerotic from the choroid the whole way round for a distance of about half an inch beyond the edge of the cornea. Make four radial cuts, equidistant from one another, through the margin of the cornea and*

the sclerotic, taking care not to injure the deeper parts; and extend the cuts back towards the optic nerve. Carefully peel off the four flaps into which the sclerotic is now divided from the underlying black choroid coat: turn them down, and pin them to the dissecting board so as to fix the eye with the iris upwards: note the following points:

i. The **ciliary muscle** is a whitish ring of unstriped muscle connecting the outer margin of the iris with the junction of the cornea and sclerotic.

ii. The **choroid** is the dense black coat exposed by the removal of the sclerotic.

iii. The **ciliary vessels** pierce the sclerotic to convey blood to and from the choroid, which is extremely vascular.

iv. The **ciliary nerves** are seen passing through the sclerotic to the choroid while the flaps are being turned down.

4. *Make a couple of radial incisions, about ½ inch apart, through the iris and ciliary muscle, and turn back the portion of the iris between the two cuts, so as to expose its hinder surface.*

i. The **ciliary processes** are a series of densely pigmented and close-set radial folds on the hinder surface of the outer margin of the iris: they fit into corresponding folds in the ligament which surrounds and supports the lens.

5. *With a large pair of scissors cut the eye into two halves by a horizontal incision at its equator. Turn over the anterior or outer half, and examine it from behind: note the following parts:*

i. The **ciliary processes**.

ii. The **uvea** is the layer of dense black pigment at the back of the iris and ciliary processes.

iii. The **ora serrata** is the indented anterior boundary of the part of the rètina sensitive to light: in front of this the retina becomes extremely thin, but really extends forwards as far as the free edge of the iris.

6. *Turn the anterior half of the eye over, so that its outer or corneal surface is uppermost: cut away the iris completely: note the following points:*

i. The capsule of the lens is an elastic transparent membrane holding the lens in its place.

ii. The suspensory ligament of the lens, or zonule of Zinn, is the outer margin of the capsule of the lens: it is marked with radiating folds into which the ciliary processes fit.

iii. The cut edges of the retina and choroid should be recognised.

7. *Remove the lens from its capsule; note its shape, more convex behind than in front: harden it with spirit, or by boiling for a few minutes in water.*

8. *Remove the vitreous humour from the posterior half of the eye: note the following points:*

i. The retina is a delicate pulpy membrane between the vitreous humour and the choroid.

ii. The blind spot is the point of entrance of the optic nerve: the retina adheres firmly to this spot, though it can be readily separated from the choroid at all other parts.

iii. The retinal vessels enter with the optic nerve, and radiate from the blind spot.

C. Histology of the Eye.

1. **The Choroid.** *Spread a small piece of fresh choroid on a slide in normal salt solution; examine with low and high powers.*

i. The choroid is a network of bloodvessels thickly invested by pigment cells.

ii. The pigment cells are irregularly branched, with clear nuclei.

2. **The Lens.** *Tease in glycerine a small piece of lens, hardened by boiling; examine with low and high powers: note:*

i. The laminated character of the lens as a whole.

ii. The elongated epithelial cells of which the lens is composed.

iii. The serrated edges of many of the cells.

3. The Retina.

Mount in balsam one of the prepared sections of the posterior part of the frog's eye: examine with low and high powers.

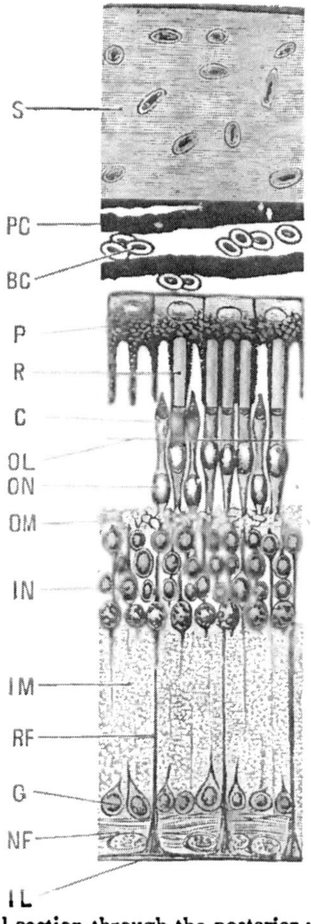

FIG. 18.—Vertical section through the posterior wall of the eye of a frog; the section passes through the sclerotic, the choroid, and the entire thickness of the retina. × 300.

BC, red blood corpuscle; C, cone; G, ganglion cell; IL, inner limiting membrane; IM, inner molecular layer; IN, inner nuclear layer; NF, layer of nerve fibres; OL, outer limiting membrane; OM, outer molecular layer; ON, outer nuclear layer; P, pigment cell of retina; PC, pigment of choroid; R, rod; RF, radial or Müller's fibre; S, the cartilaginous sclerotic.

a. The sclerotic consists chiefly of hyaline cartilage.

b. The **choroid** is a vascular plexus, with much pigment.

c. The **retina** is composed of the following layers from without inwards:

i. A layer of **pigment cells**, sending processes between the rods and cones.

ii. The **rods and cones** are a layer of columnar bodies placed vertically to the surface: the rods are far more numerous, and much larger than the cones: each consists of an inner and an outer segment.

iii. The **outer limiting membrane** is seen in sections as an exceedingly thin line separating the rods and cones from the outer nuclear layer.

iv. The **outer nuclear layer** is a moderately thick layer, well stained, and very closely connected with the rods and cones.

v. The **outer molecular layer** is a very thin layer, not stained: it is finely granular or reticular in structure.

vi. The **inner nuclear layer** is thick and well stained. It consists of several layers of large nuclei, surrounded by thin layers of protoplasm, produced into radial processes.

vii. The **inner molecular layer** is a thick, finely granular layer, not stained.

viii. The layer of nucleated **nerve cells** (ganglion cells) is a single layer of large branched cells.

ix. The layer of **nerve fibres** is formed by the branches of the optic nerve.

x. The **inner limiting membrane**, formed by the cone-shaped ends of Müller's fibres.

xi. The **radial fibres**, or **Müller's fibres**, commence with expanded ends in the layer of nerve fibres and stretch outwards: they can easily be traced through the inner molecular layer. They consist of connective tissue and serve to strengthen and support the retina.

D. The Ear of the Frog.

The frog's auditory organ is too small to dissect satisfactorily, and is best studied by making transverse sections of the entire head in the following manner:

Kill a frog with chloroform; cut off the head with stout scissors, and decalcify it by placing in a 5 to 10 per cent. solution of nitric acid, or in a mixture of chromic acid with a few drops of nitric acid. When the bones are thoroughly soft, which will take from a few hours to 3 or 4 days or more according to the strength of acid employed, remove the head from the decalcifying solution and transfer to weak alcohol and thence to strong alcohol. Then stain with borax carmine and imbed in paraffin, and cut into transverse sections with a microtome. Mount the sections in series; examine and draw them, showing the following points:

1. The **periotic capsule** consists mainly of cartilage, and is firmly fused with the hinder part of the cranium.
2. The **vestibule** is a membranous sac lying in the cavity of the periotic capsule, and filled with a watery fluid, the **endolymph**: it is partly divided by a constriction into two main divisions:
   - i. The **utriculus** is the upper and larger division.
   - ii. The **sacculus** is the inferior and smaller division: from it arise three small saccular dilatations, supposed to represent the cochlea of higher animals.
   - iii. The **ductus endolymphaticus** rises from the inner

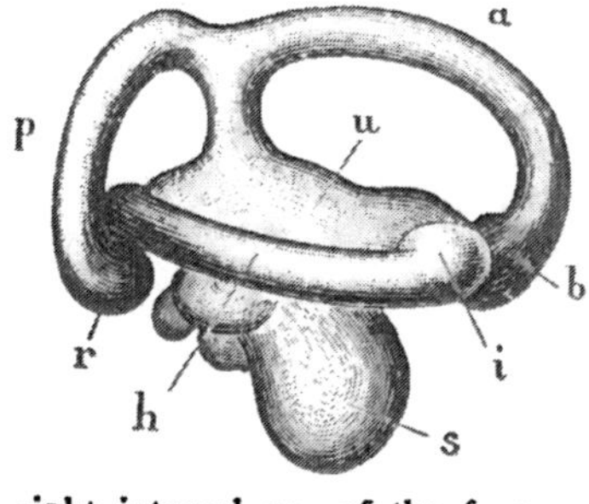

FIG. 19.—The right internal ear of the frog, removed from the periotic cartilage and drawn from the outer surface.

a, the anterior vertical semicircular canal; *b*, its ampulla; *h*, the horizontal canal; *i*, its ampulla; *p*, the posterior vertical canal; *r*, its ampulla; *s*, the sacculus; *u*, the utriculus.

and upper border of the sacculus as a tube which terminates in a large thin-walled sac.

3. The **semicircular canals** are three tubular offsets of the vestibule, into which they open at both ends. They lie in canals in the periotic cartilage, and are placed in planes at right angles to one another; each has at one end, close to its opening into the vestibule, a dilatation or **ampulla**.
   - i. The **anterior vertical canal** has its ampulla at its anterior end.
   - ii. The **posterior vertical canal** has its ampulla at its outer end, while its inner end joins the posterior end of the anterior vertical canal to open into the vestibule by a common orifice.
   - iii. The **horizontal** or **external canal** has its ampulla at the anterior end.

4. The **auditory nerve** leaves the cranial cavity through a hole in the inner wall of the periotic capsule, and divides into branches distributed to the sacculus and its diverticula, and to the ampullæ of the semicircular canals.

5. The **accessory auditory apparatus.** The essential organs of hearing—*i.e.*, the vestibule and its offsets, and the auditory nerve—are enclosed in the periotic cartilage, which is deeply placed in the side of the head: the communication with the surface is brought about by the accessory apparatus, which consists of the following parts (Fig. 10, p. 48).
   - i. The **Eustachian passage** and **tympanic cavity** are formed in connection with the hyomandibular gill cleft of the tadpole. (See Chap. VIII., p. 126.)
   - ii. The **tympanic membrane** closes the tympanic cavity on its outer side.
   - iii. The **columella** is a rod of bone and cartilage, the outer end of which is attached to the tympanic membrane, while its inner end is inserted into a hole in the outer wall of the periotic capsule (the fossa fenestræ ovalis) so as to lie in close contact with the vestibule. It serves to communicate the auditory vibrations of the tympanic membrane to the vestibule.

CHAPTER VII.

THE REPRODUCTIVE ORGANS AND THE CLOACA OF THE FROG.

A. The Male Frog.

1. The Reproductive Organs. (Fig. 3, p. 20.)

Pin the frog on its back under water; open the body cavity from the ventral surface; turn aside the alimentary canal and liver.

i. The **testes** are a pair of yellow ovoid bodies about half an inch long, lying on the ventral surface of the kidneys. Within the testes are developed the essential male elements or spermatozoa.

ii. The **vasa efferentia** are a number—usually 10 to 12—of slender ducts, connecting the testis of each side with the inner or median border of the corresponding kidney; they serve to convey the spermatozoa from the testis into the tubules of the kidney, whence they escape by the ureter, which acts as vas deferens.

iii. The **vas deferens** or **ureter** runs along the outer side of the posterior part of the kidney, and then backwards to the cloaca.

iv. The **vesicula seminalis** is a large pouch-like dilatation on the outer side of the vas deferens, just behind the kidney and before reaching the cloaca.

2. The Cloaca. (*Cf.* Fig. 3, p. 20.)

Lay the frog on its back; with a stout scalpel split the pelvic symphysis in the median plane; gently separate the two halves, and pin them out right and left, so as to expose the cloaca from the ventral surface.

i. Into the cloaca open the large intestine, the ureters, and the bladder.

ii. The **bladder** is a thin-walled bilobed muscular sac, lying on the ventral surface of the large intestine and cloaca, its two lobes communicating freely with each other. It is invested by peritoneum and attached to the sides of the body by special peritoneal folds.

Inflate the bladder with a blow pipe inserted through the cloacal aperture: pass a seeker up the cloaca to determine the exact position of the opening from the bladder to the cloaca. Cut up the cloaca along one side; wash out its contents and examine the opening into the bladder.

iii. The **ureter** or **vas deferens** is continued behind the vesicula seminalis as a very short tube, opening into the dorsal wall of the cloaca almost exactly opposite the opening of the bladder on the ventral surface. The openings of the two ureters are close together on the apices of two small papillæ, overhung by a slight valvular projection of the mucous membrane of the cloaca.

iv. *Remove a small piece of the testis; place it on a slide in a drop of salt solution; press it slightly; cover and examine with a high power to see the spermatozoa.*

B. The Female Frog.

1. The Reproductive organs.

Dissect as in the male.

i. The **ovaries** are a pair of black masses lying in folds of the peritoneum ventral to the kidneys, in very much the same position as the testes in the male. Their shape, colour and size vary much at different seasons of the year. On their surfaces are numerous rounded projections, like small shot; these are ova in various stages of development; the smaller and younger ones are white; the larger and more mature ones black in one half and white or yellowish in the other. Each ovary consists of a couple of folds united along their ventral edges; the space between the folds is divided by partitions into about fifteen pouches.

ii. The **oviducts** are a pair of white, twisted tubes, with thick gelatinous walls. They commence with open mouths at the extreme front end of the body cavity, close to the outer side of the roots of the lungs; and run back, increasing in size, and becoming much convoluted. Their hinder ends, the **ovisacs**, are greatly dilated, but have thinner walls; in these the eggs are stored just before laying commences. Unlike the male, the female has genital ducts distinct from the ureters.

2. The Cloaca.

Dissect as in the male.

i. The **cloaca** is very similar to that of the male, except that the urinary and genital products are discharged into it by separate ducts.

ii. The **bladder** is like that of the male.

iii. The **oviducts** open separately into the dorsal wall of the cloaca, just opposite the bladder, by two wide apertures separated by a narrow median partition.

iv. The **ureters** open by two small apertures, placed close together, into the dorsal wall of the cloaca just behind the oviducts.

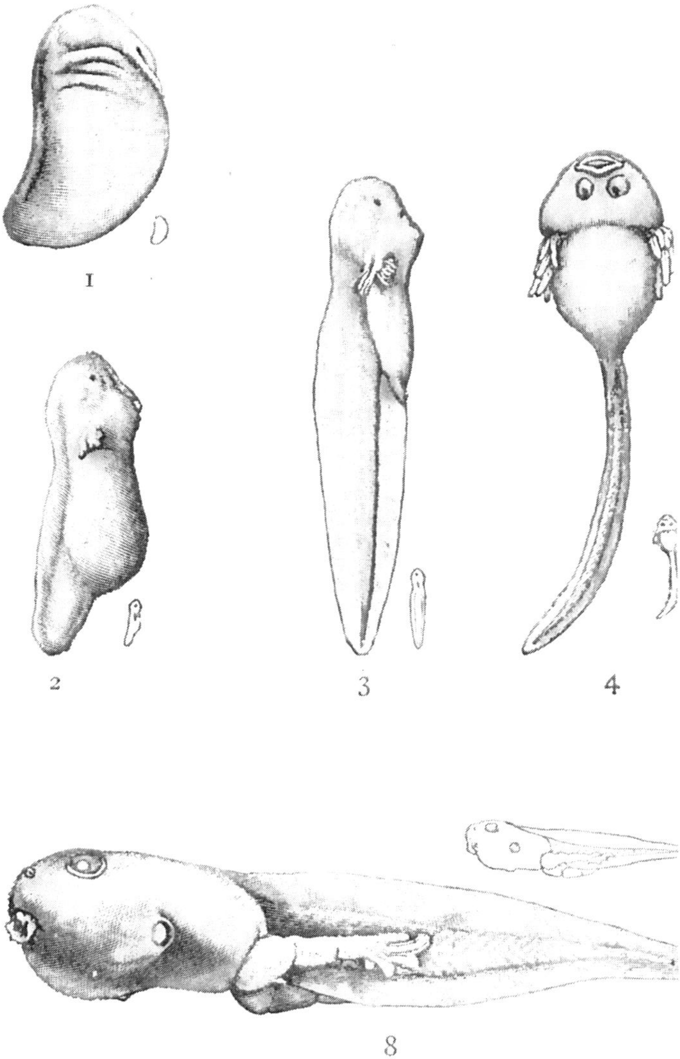

FIG. 20.—STAGES IN THE DEVELOPMENT OF THE FROG.

1. Embryo from the right side, showing the ventral gland, the hyo cleft and of three gill clefts : a rounded swelling anteriorly marks the s showing the rudiments of the external gills. 3. Larva from the right below, showing the mouth and horny jaws, the ventral gland split int 5. From below : the operculum has covered the gills except on the left opercular cavity, and rudiment of hind limb just above the rectum. anus. 8. From the left side, at the commencement of metamorphosis parent operculum, the elbow projecting. 9. Young tailed frog.

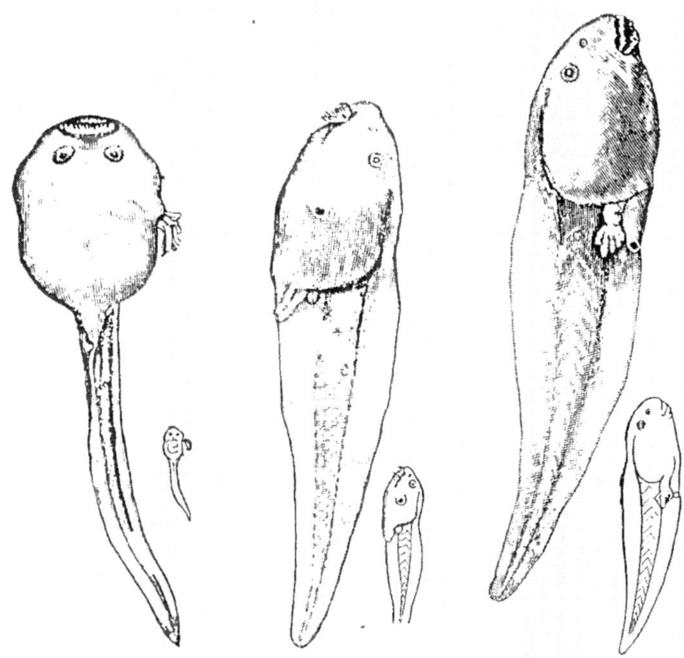

5 6 7

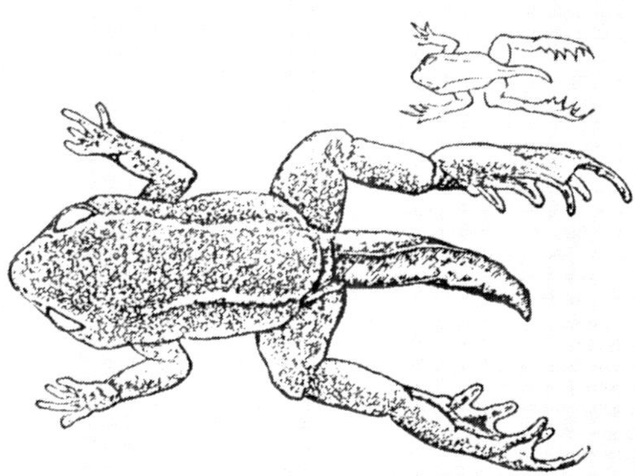

9

actual sizes are shown by the small outline drawings.)

and two branchial arches, and indications of the hyomandibular
she eye. 2. Embryo at the time of hatching, from the right side,
with well developed external gills, nose, and eye. 4. Larva from
the transverse fold indicates the rudiment of the operculum.
6. From the left side: showing the beak, nose, eye, opening of
om the right side, showing the developing hind limb, and the
tail beginning to shorten: the fore limb seen through the trans-

Between pages 98 and 99.

CHAPTER VIII.

THE DEVELOPMENT OF THE FROG.

I. General Account.

The frog's eggs are laid in water, usually during March or the early part of April.

During the act of oviposition, which may last several days, the male frog clasps the female firmly, embracing her with his arms; and as the eggs pass out from the cloaca of the female they are fertilised by spermatozoa discharged over them by the male.

The eggs, which are very numerous, are small spherical bodies about 1·75 mm. in diameter; they are invested by thin coatings of an albuminous substance, which swell up very greatly in the water, and stick together to form the bulky masses we call frog's spawn. Such spawn consists of a transparent gelatinous mass, formed by the swollen albuminous matter, in which are embedded the eggs: these latter appear as small round bodies, each presenting a black half and a white half.

If a number of hen's eggs were broken into a basin, care being taken not to rupture the yolks, a mass would be produced similar to frog's spawn: the yellow yolks corresponding to the frog's eggs, and the whites or albuminous investments of the yolks to the gelatinous matrix of the spawn. And just as the chicken is formed from the yolk, and not from the white of a hen's egg, so also is the frog developed from the egg and not from the gelatinous investment.

The frog's eggs, laid in this way and fertilised by the spermatozoa shed over them by the male, begin to develop at once. Each egg is at first spherical, but in about a week becomes ovoid in shape, and then rapidly increases in length. By the tenth day it is divided by slight constrictions into head, body, and tail. The whole animal becomes fish-like in appearance, the tail growing rapidly: two pairs of branching tufts, the external gills, followed shortly by a third pair, grow out from the sides of the neck, and in about a fortnight from the time of laying of the eggs the young tadpoles make their way out of the gelatinous mass of the spawn, and swim freely in the water.

The egg is surrounded by a tough membrane through which the tadpole has to find a way. The method of hatching appears to be as follows. The tadpole possesses a frontal gland immediately in front of the horseshoe-shaped cement-organ on the under side of its head. Fixing this cement-organ to one point of the egg-membrane, it performs movements around this fixed point by the aid of the cilia which cover the greater part of its body, and rubs the frontal gland over this part of the vitelline membrane. Ultimately the membrane is softened and gives way, allowing the tadpole to creep out and adhere to weeds.

At the time of hatching the tadpole has no mouth, and is dependent for food on granules of food-yolk which are contained in large numbers in the egg, and at the expense of which all the earlier processes of development are effected.

A few days after hatching the mouth appears, bordered by a pair of horny jaws, and fringed with fleshy lips provided with horny papillæ. The alimentary canal which has hitherto been wide and short, now rapidly increases in length, becoming tubular and convoluted in the form of a reversed spiral; the liver and pancreas are formed; the anus is developed even before the mouth, and the tadpole now feeds eagerly on confervæ and other vegetable matter.

About the time of appearance of the mouth, *i.e.*, shortly after hatching, a series of four slit-like openings, the gill clefts, appear on each side of the neck, leading from the pharynx to the exterior. The ectoderm covering them becomes tasselled and forms a series of opercular gills. At the same time the earlier formed more dorsally placed external gills degenerate.

While external gills bordering the gill-slits are developing, a fold of skin, the operculum, appears on each side of the head in front of the gills, and grows backwards over these, so as to enclose them in a gill chamber. Towards the end of the fourth week the hinder edges of the opercular folds fuse

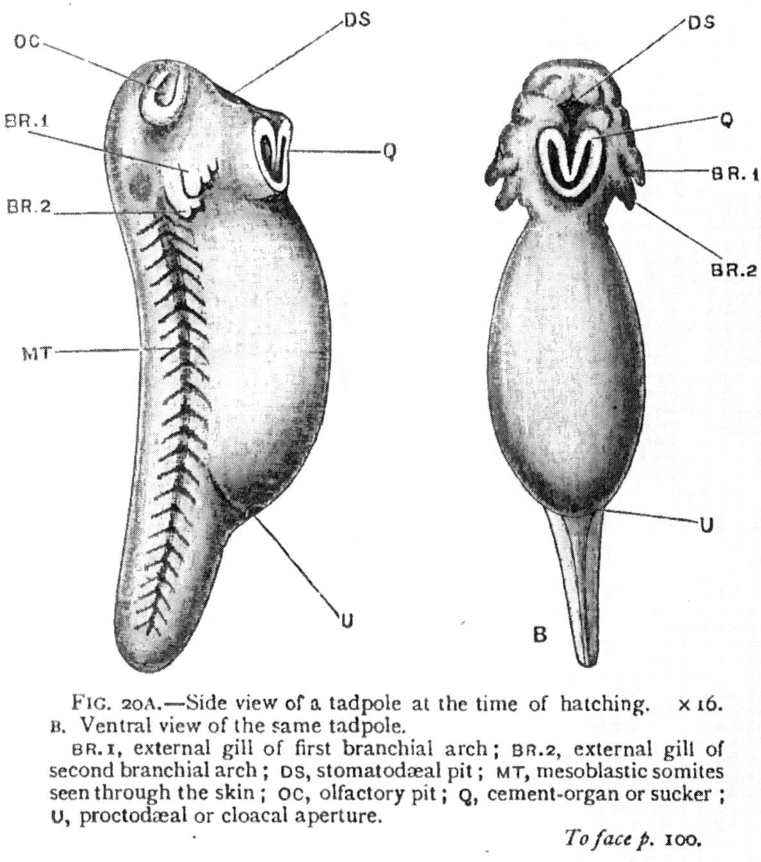

FIG. 20A.—Side view of a tadpole at the time of hatching. ×16. B. Ventral view of the same tadpole.

BR.1, external gill of first branchial arch; BR.2, external gill of second branchial arch; DS, stomatodæal pit; MT, mesoblastic somites seen through the skin; OC, olfactory pit; Q, cement-organ or sucker; U, proctodæal or cloacal aperture.

To face p. 100.

with the body-wall on the ventral surface and along the right side. On the left side a spout-like opening remains which communicates with the gill-chambers of both sides, and through which the water taken in at the mouth for respiration, and passed through the gill-slits, makes its escape to the exterior, Fig. 20, $_{5}$, $_{6}$, $_{8}$. It is the inclusion of the gills by this opercular fold that has given rise to the notion of opercular or "internal" gills. True internal gills, however, are not present in the frog or its allies at any stage of development.

During this time the tadpole has been feeding freely, and has increased greatly in size. The body, Fig. 20, $_{5}$, is broad and round; the tail is much larger than before, and forms a powerful swimming organ; while the cement-organ on the under surface of the head, though still present, is small and but little used for adhesion.

Very shortly afterwards rudiments of the hind limbs can be seen as a pair of small papillæ at the root of the tail, one on each side of the anus, Fig. 20, $_{6}$; these steadily increase in size: about the seventh week they become divided into joints; and a week later the toes appear, Fig. 20, $_{7}$.

The fore-limbs arise about the same time as the hind ones, but are covered by the opercular folds, and hence do not become visible till a later stage, Fig. 20, $_{8}$.

Towards the end of the second month the lungs come into use, and the tadpoles which now have the form shown in Fig. 20, $_{8}$, frequently come to the surface of the water to breathe. The gills now begin to degenerate, but for a time respiration is effected both by the gills and the lungs.

A fortnight or three weeks later a distinct metamorphosis occurs, whereby the tadpole becomes transformed from the fish-like condition in which it has hitherto remained to the purely air-breathing stage characteristic of the adult. The tadpole ceases to feed; a casting, or ecdysis, of the outer layer of the skin takes place; the gills are gradually absorbed; the horny jaws are thrown off; the large frilled lips shrink up; the mouth loses its rounded suctorial form and becomes much wider; the tongue, previously small, increases considerably in size; the eyes become larger and more prominent;

the fore-limbs appear, the left one being pushed through the spout-like opening of the branchial chamber, and the right one forcing its way through the opercular fold, in which it leaves a ragged hole. The abdomen shrinks; the stomach and liver enlarge, but the intestine becomes considerably shorter than before, and of smaller diameter; the animal, previously a vegetable feeder, now becomes carnivorous. The gill-clefts close up; and important modifications accompanying the change in breathing, occur in the blood-vessels.

The tail, which is still of great length, Fig. 20, $_8$, now begins to shorten, and is soon completely absorbed; the hind-limbs lengthen considerably, and the animal leaves the water as a Frog, Fig. 20.

Explanation of the Metamorphosis.

The structure and mode of life of the tadpole (opercular gills; gill slits; a tail; and lateral sense organs) give it a fish-like character. But there is no known fish which at all closely resembles the tadpole. The mud-fish of Africa and America approach it. Still more close is the agreement between the tadpole and the young stage of these fish. Hence we may say that the most probable explanation of the problem, "Why the frog first becomes a tadpole?" is that its fish-like ancestors passed through a similar stage in their career. The frog, in brief, climbs up its genealogical tree in its own life-history.

Simplification of the Frog's Structure.

Before following the development of the frog in detail it will be useful if we imagine its structure to be reduced to the most essential features; and anticipating one or two of the developmental steps, trace in thought the simplification of these essential organs to the earlier stages and ultimately to the egg.

From this point of view we may look upon the frog as simply a trunk, disregarding the head and limbs. This step is justified by the structure of the lowest vertebrates in which no limbs are present and only a very indistinct head is developed. Essentially, then, the frog consists of a muscular body-wall forming a tube thick above, thin below, and bounding the body-cavity. On the dorsal surface is a tube, the central nervous system, imbedded in the body-wall. In the body-cavity is a second tube, the alimentary canal. Around the nervous tube is a bony sheath divisible into skull and vertebræ. The most essential part of this skeleton is a rod of cells running down the centra. In the lowest vertebrates this alone persists and is called the **notochord.** Thus reduced to its essential anatomical characters, a frog consists of two tubes lying one over the other, separated by a rod, the whole enclosed in a flexible muscular body-wall.

The wall of the alimentary tube and that of the body are made up of three complementary layers. Counting from the body-cavity inwards to the gut-cavity, the coats are peritoneum, muscle, and mucous lining: from the body cavity outwards they are peritoneum, muscle, and mucous skin.

The development of the frog is mainly the history of the formation of these two tubes, one nervous the other digestive; of the intervening rod or notochord; and of the coats which compose the body-wall and the gut-wall.

We may anticipate the first history. The nervous tube is derived from the skin. It is folded off along the back, surrounded by nerve-cells and sunk inwards. We are thus left with a double tube: the alimentary canal and the body-wall with its notochordal rod.

The complementary nature of the inner and outer coats of the body-cavity suggest that they have a common origin in a common tissue. This would reduce the frog to a tube composed of a central digestive layer surrounded by a middle coat of indifferent tissue and covered with the outer skin. These three coats have received names. The outer one, being potentially both nervous system and skin, is called by a name signifying surface-growth or **epiblast**; the middle one, **mesoblast**; the inner, **hypoblast**.

There are two ways in either of which this triple-layered organism could arise from a simple beginning. Imagine a hollow ball pushed in at one pole while the space between the inner and outer cavities was filled up by growth of tissue round their junction. Or, starting with a solid ball, imagine a liquefaction to occur at the centre, an inner layer of tissue condensing round this central cavity and round the outer surface. We must now anticipate that the latter process is the one chiefly followed by the frog, and that it starts at the lower pole of the ball. Thus we come to the conception of the triple-layered embryo arising from a solid or nearly solid mass of cells whose upper and lower poles differ: and from this it is an easy step to the next, a single cell with different poles. This cell is the egg, whose fortunes we shall now follow in the reverse direction.

II. Detailed Account.

A. Formation of the Egg.

The early stages in the formation of the egg cannot be seen in the adult frog, but must be studied in tadpoles. In tadpoles of about 10 mm. length, shortly after the opening of the mouth, a pair of longitudinal ridge-like thickenings of peritoneum appear along the dorsal surface of the body cavity, close to the mesentery, and along the inner borders of the developing kidneys. These **genital ridges** are found in all tadpoles of this age, no distinction of sex appearing until a much later period.

Each genital ridge is at first due merely to slight modification in the shape of the peritoneal epithelial cells which, elsewhere flattened, become here cubical or slightly columnar. The ridges soon become more prominent, especially at their anterior ends; their growth being due partly to the epithelial cells increasing by division so as to form a layer several cells thick, and partly to the ingrowth of an axial core of connective tissue from the basal membrane of the peritoneum along which the bloodvessels gain access to the ridge.

From the posterior two-thirds of the genital ridge the ovary, or in the male the testis, is developed; while the anterior third undergoes degenerative changes, and becomes converted into the fat body.

The Primitive Ova. At an early stage certain of the epithelial cells of the genital ridge become conspicuous by their larger size and more spherical shape; and around these larger cells, or **primitive ova,** as they are called, the smaller epithelial cells become arranged so as to form capsules or follicles; the follicles, with their contained primitive ova, form small knob-like projections on the surface of the genital ridge. New primitive ova arise either directly from the surface epithelium, or by division of the already existing ones.

The Permanent Ova. Up to this time there has been no distinction between male and female, the processes described occurring in all tadpoles alike. Sexual differentiation appears about the time of the metamorphosis. In the female the change consists essentially in a great increase in the size of the genital ridge, which now becomes the ovary, and in the formation of the **permanent ova,** or eggs. The permanent ova are derived from the primitive ova; in some cases each primitive ovum is directly converted into a permanent ovum, but it has been suggested that in others two or more primitive ova are concerned in the formation of a single permanent ovum.

A permanent ovum is enclosed in a follicle or capsule like the primitive ovum, and differs from this latter in the following points: (1) it is of larger size; (2) it contains within its substance a number of small sharply defined yellowish granules of food yolk, which are elaborated by the follicle cells and passed on from them to the ovum; these yolk granules increase rapidly in number, and to them the greater size and opacity of the permanent ovum are chiefly due; (3) important changes have occurred in the nucleus; in the primitive ovum the nucleus is small, granular in appearance, and apparently solid; in the permanent ovum the nucleus, or **germinal vesicle,** is of very large size, up to half the diameter of the entire ovum, and consists of an elastic capsule or nuclear membrane, filled with fluid and traversed by a protoplasmic reticulum enlarged at its nodes to form the nucleoli, or **germinal spots.**

When the permanent ovum has reached a diameter of about 0·5 mm., an exceedingly thin structureless investment, the vitelline membrane, is formed immediately around it, within the follicle. The mode of origin of the vitelline membrane is not clearly made out, but it seems to be formed from the ovum itself rather than from the follicular epithelium.

A little later still a layer of black pigment appears on the surface of the ovum; it is at first irregularly distributed over the whole surface, but as the ovum ripens it becomes restricted to one half or hemisphere. The pigment is contained, and apparently formed, within the ovum itself, but it is not clear how it is formed or what purpose it fulfils.

B. Maturation of the Egg.

The eggs have now reached their full size, and project from the surface of the ovary like small shot; but they have still to pass through the process of **maturation**, or ripening, before they are ready to be fertilised. This process of maturation concerns the nucleus almost exclusively.

The nucleus, which at its full size we have seen to be quite half the diameter of the egg itself, begins to shrink; the nuclear membrane becomes wrinkled, its surface presenting a number of small wart-like projections, so that the whole nucleus has a blackberry-like appearance. Part of the nuclear fluid exudes through the nuclear membrane into the substance of the egg; a great part of the nuclear reticulum disappears, or becomes broken up into isolated globules or nucleoli, but a very small part remains in the centre as a slender intricately coiled thread, the **nuclear skein**.

About this time the eggs are discharged from the ovary, the follicles rupturing, and the eggs falling into the body cavity of the frog; along this they pass forwards, directed partly by contraction of the muscular body-walls, partly by the action of the cilia of the peritoneum, to the mouths of the oviducts, which are situated at the anterior end of the body cavity opposite the roots of the lungs. In the first, or thick-walled part of the oviduct the eggs acquire gelatinous investments, secreted by glands in its walls. The terminal, or hinder, part of the oviduct forms a thin-walled sac capable of great distension, within which the eggs accumulate in large numbers. Finally, the eggs are passed out through the cloaca into water, in which the albuminous investments of the eggs speedily swell up to form the gelatinous mass of the frog's spawn.

During the discharge of the egg from the ovary, and its passage down the oviduct, further changes occur in its nucleus. The nuclear membrane still further collapses, and finally disappears completely; the nuclear fluid and nucleoli become distributed through the substance of the egg, and of the original large nucleus the exceedingly minute nuclear skein alone remains.

This nuclear skein moves from the centre of the egg to its surface, which it reaches opposite the centre of the black hemisphere. The skein, previously an irregularly tangled thread, now assumes the definite form and arrangement of a **nuclear spindle**, such as may be seen in the nucleus of an epithelial or other cell immediately before division of the cell occurs.

The First Polar Body. About the time the egg is laid, but before it is fertilised, the egg becomes slightly flattened at its upper or black pole, a certain amount of fluid being exuded between the egg and the vitelline membrane. The nuclear spindle now divides into two equal parts, one of which remains within the egg, while the other is extruded from it as the **first polar body**, a minute ovoidal white globule, which lies on the surface of the egg in the exuded peri-vitelline fluid.

The Second Polar Body. The half of the nuclear spindle that remains within the egg retreats from the surface a little distance, and then divides into two equal parts, one of which remains within the egg as the **female pronucleus**, while the other is extruded as the **second polar body**, a minute white globule very similar to the first polar body, and like this lying freely in the perivitelline fluid on the top of the egg.

In the case of most animals in which the formation of polar bodies has been observed, both the first and second polar bodies are extruded before fertilisation is effected. In the frog the extrusion of the second polar body does not occur until after the spermatozoon has entered the egg, though before the completion of the act of fertilisation.

C. Fertilisation of the Egg.

Fertilisation, or impregnation, consists in fusion of the spermatozoon with the egg; or, more strictly speaking, fusion of the nuclei of these two bodies.

The spermatozoa, after being shed over the spawn by the male, swim actively by means of their long tails, penetrate the gelatinous investment of the eggs, bore their way through the vitelline membrane, and so penetrate into the eggs themselves, which they enter at or close to the upper or black pole.

A single spermatozoon is sufficient to fertilise an egg, and it is doubtful whether more than one is ever normally concerned in the process.

In about an hour after the spermatozoon has entered, a pigmented process may be seen projecting inwards from the surface of the egg, with a clear spot in its centre. This spot is the nucleus of the spermatozoon, and is spoken of as the **male pronucleus**: it penetrates farther into the egg, carrying the pigment with it, so that it appears surrounded by a pigmented capsule connected with the surface of the egg by a pigmented stalk.

By this time the second polar body has been formed and extruded, and the **female pronucleus** is the only part of the original egg nucleus still remaining. The male and female pronuclei, which are at first some little distance apart, rapidly approach each other, come into close contact, and after having increased considerably in size, fuse together, about two and a half hours after fertilisation has commenced, to form the **segmentation nucleus.**

The segmentation nucleus is a large spherical vesicle imbedded in finely granular protoplasm, and surrounded by an ill-defined capsule of pigment: its formation by the fusion of the male and female pronuclei completes the act of fertilisation.

The female pronucleus may be regarded as an imperfect nucleus, and the upshot of the process of fertilisation is the completion of this nucleus; the nucleus of the spermatozoon, or male pronucleus, replacing the part of the egg-nucleus which has been lost as the polar bodies. The further explanation of the sexual process is probably to be found in the great advantage, as regards vigour of offspring, that is known to result in both animal and vegetable kingdoms from cross fertilisation, *i.e.*, from combining the protoplasm of two distinct individuals in the act of reproduction.

D. Segmentation of the Egg.

The earliest stages of development consist in repeated division of the egg, whereby it becomes converted from the unicellular condition, which is permanent only in the lowest animals, to the multicellular state characteristic of all higher animals. To these early processes of development the name **segmentation** is given.

Very shortly after the completion of the act of fertilisation and formation of the segmentation nucleus this latter loses its spherical form and becomes spindle-shaped, the yolk granules at the same time showing a tendency to become arranged in lines radiating outwards from the ends of the spindle. The nucleus now divides into two halves, which move away from each other; the yolk granules tend to aggregate themselves around the two nuclei, and a thin vertical plate of finely granular substance is left, bisecting the egg.

At the upper or black pole of the egg a depression now appears, at first as a small pit and then as a groove, which soon extends all round, and, rapidly deepening, divides the egg into two completely separate halves along a plane corresponding with the vertical plate mentioned above.

Each of the two nuclei soon divides again into two, and a second cleft is formed in the same manner as before: it also is vertical, but in a plane at right angles to the first one, and on its completion the egg consists of four precisely similar segments or cells, each with a nucleus.

The third cleft is horizontal, but not equatorial, lying nearer the upper than the lower pole: it divides each of the four cells into two, an upper smaller and a lower larger one.

Two more vertical clefts next appear simultaneously at the upper pole, midway between the two primary clefts, and extending downwards divide first the smaller and then the larger cells, giving sixteen cells in all, eight smaller upper ones, and eight larger lower ones. Two more horizontal clefts then appear, which again double the number of segments, giving thirty-two in all.

From this stage segmentation proceeds in a less regular manner, the upper and smaller cells dividing more rapidly than the lower and larger ones. By means of radial and concentric

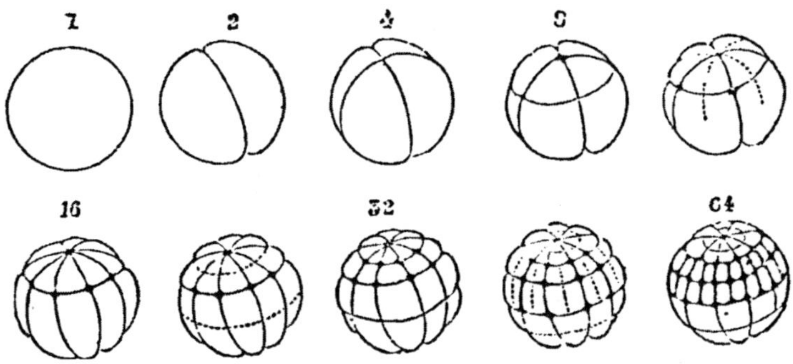

FIG. 21.—Segmentation of the frog's egg. (From Haddon, after Ecker.)

The numbers above the figures indicate the number of segments at the several stages. The dotted lines mark the positions of the clefts that will next appear.

clefts, the number of cells is rapidly increased, division of the cells being in all cases, as from the first, preceded by division of their nuclei.

At the stage when only eight cells are present, *i.e.*, on the completion of the third cleft, a small cavity appears in the centre of the egg, round which the cells are grouped: during the later phases of segmentation this **segmentation cavity** or **blastocoele**, as it is called, increases considerably in size: it is from the first situated nearer the upper than the lower pole of the egg, and is filled with fluid.

At the close of segmentation the egg has the structure shown in section in Fig. 22.

It is a hollow ball with its walls composed of three or four layers of cells, and of very unequal thickness, owing to the segmentation cavity lying in the upper half of the egg. The cells of the upper half are small, fairly uniform in size, and regularly arranged, while those of the lower half are larger, and more irregular both in shape and size. The superficial cells of the upper half are deeply pigmented, while the cells of the lower half are almost colourless.

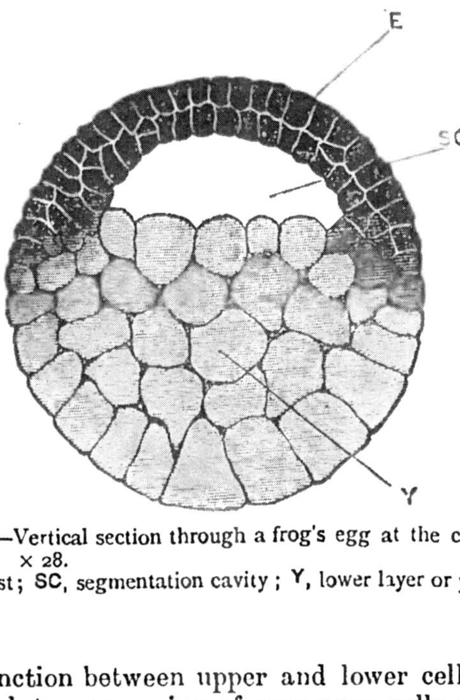

FIG. 22.—Vertical section through a frog's egg at the close of segmentation. × 28.
E, epiblast; SC, segmentation cavity; Y, lower layer or yolk cells.

The distinction between upper and lower cells is however not an absolute one, a ring of numerous cells more or less intermediate in size, shape, and depth of pigmentation, occurring round the equator of the egg at the junction of its upper and lower halves. These intermediate cells (Fig. 22A, IC) take an important part in the formation of the frog's tissues.

The process of segmentation is, as mentioned above, simply one of cell-division; and the unequal rates at which the different parts of the egg segment, are to be regarded as due to the retarding influence of the granules of food-yolk, which, being themselves inert, must hinder the activity of the protoplasm in which they are imbedded. These granules of food-yolk are more abundant in the lower than the upper half of the egg, and this unequal distribution of food-yolk is the direct cause of the unequal segmentation of the egg. The purpose of food-yolk is to afford a supply of nutriment at the expense of which the earlier developmental processes

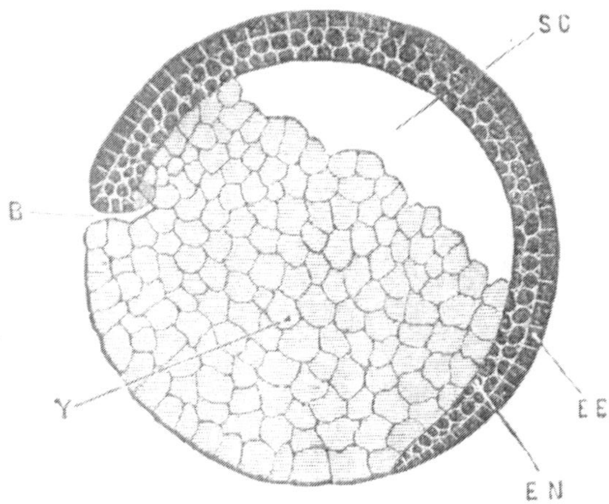

FIG. 23.—Longitudinal vertical section of a frog embryo, showing commencing hypoblast at B. ×28.

B, blastopore; EE, outer or epidermic layer of epiblast; EN, inner or nervous layer of epiblast; SC, segmentation cavity; Y, yolk-cells.

may be accomplished, until the young animal is sufficiently advanced to obtain food for itself; and the direct influence of this food-yolk will be to hinder rather than to help these processes.

E. The Germinal Layers.

At the close of segmentation we have seen that the egg consists of cells of two kinds; firstly, those of the upper half of the egg, which are smaller, pigmented, more regularly arranged, and comparatively free from food-yolk; secondly,

Fig. 22 A

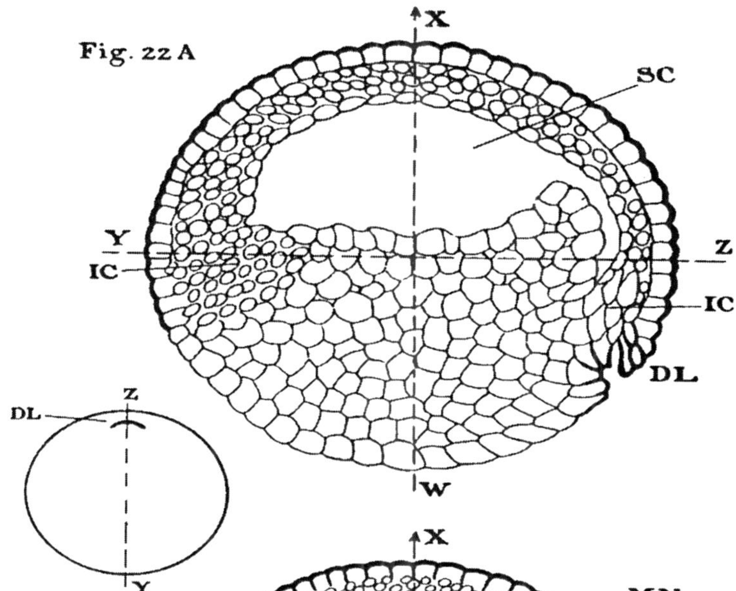

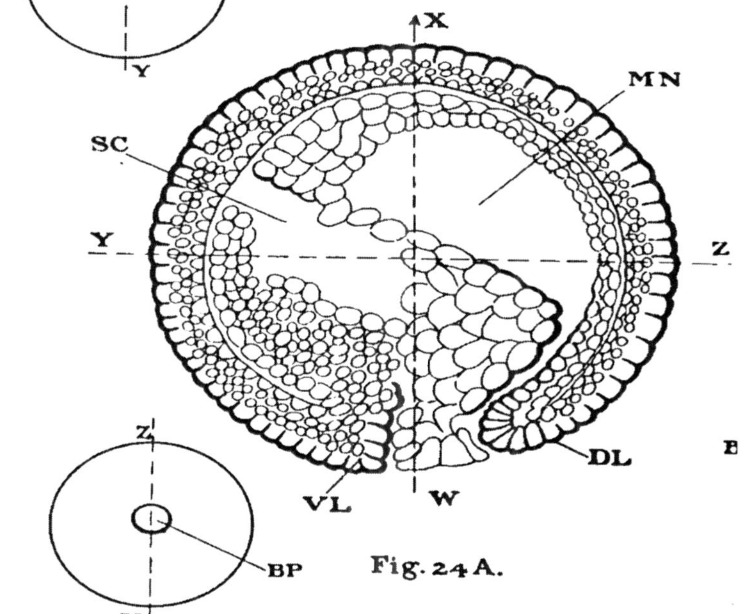

Fig. 24 A.

FIGS. 22A–25A.—The four small figures represent the the formation and shifting of the blastopore (BP). The through its horizontal axis YZ, and represent four stages of the egg through more than 90°.

DL, dorsal lip of blastopore ; IC, intermediate cells ; VL, ventral lip of blastopore ; WX, axis of embryo, whis YZ, axis of embryo, which is at first horizontal and subse

Fig. 23A.

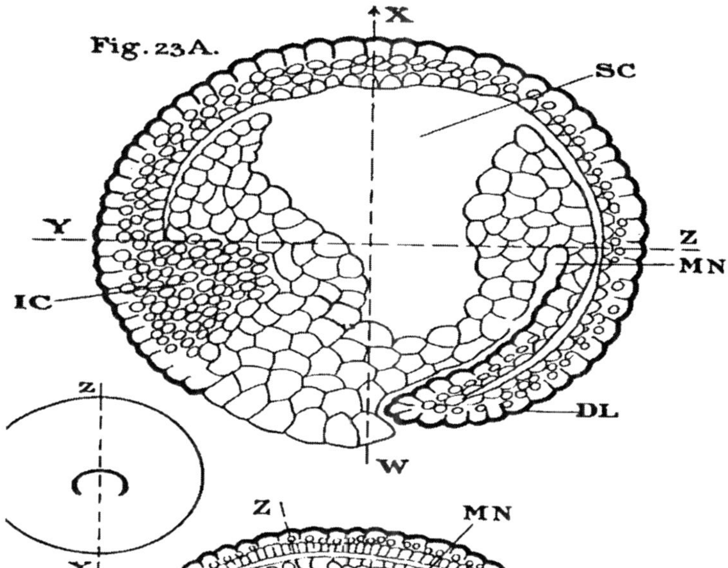

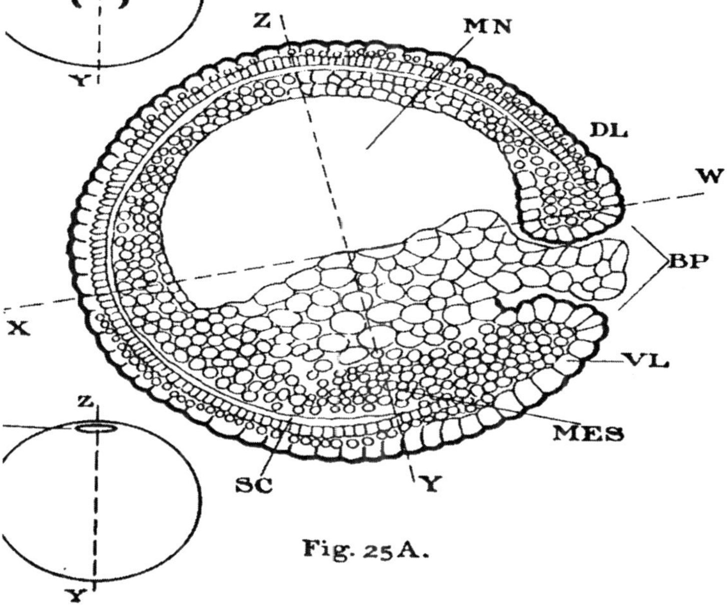

Fig. 25A.

egmenting egg, seen from its lower pole, and illustrate
our large figures are sections of the egg taken vertically
the formation of the mesenteron. Notice the rotation

IN, mesenteron ; SC, segmentation cavity or blastocœl ;
a is at first vertical and subsequently antero-posterior ;
uently vertical or nearly so.

Between pages 112 *and* 113.

those of the lower half of the egg which are larger, less regular, and almost free from pigment, but much distended by food-yolk, which is present in such quantity as to render them comparatively inert.

The former are the **epiblast cells**; the latter may conveniently be spoken of as the **lower layer cells** or **yolk-cells.**

The epiblast shows almost from the first a distinction into

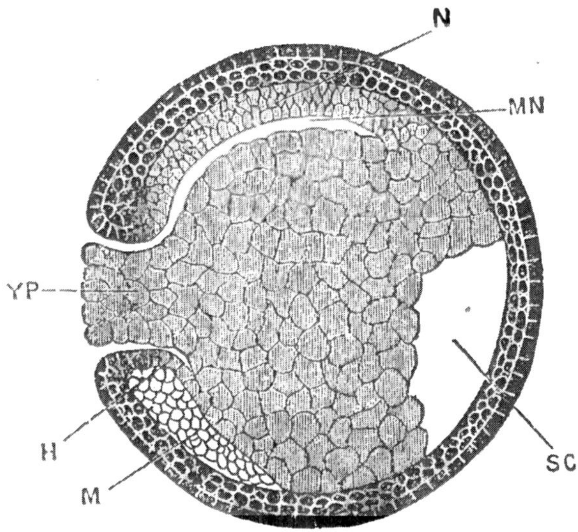

FIG. 24.—Longitudinal vertical section through a frog embryo at a later stage in the formation of the mesenteron.
H, invaginate hypoblast; M, mesoblast; MN, mesenteron; N, notochord; SC, segmentation cavity; YP, yolk plug, filling up the blastopore.

two layers; the most superficial cells being somewhat cubical in shape and closely applied side by side so as to form a continuous and deeply pigmented layer; while the deeper cells are more spherical, less strongly pigmented, and loosely arranged in a layer two or more cells deep.

The epiblast cells continue to increase by division, and very early, owing apparently to multiplication of the cells at the margin of the layer, seem to spread over the lower or yolk-cells. Owing to the difference in colour of the two halves of the egg, the various stages of this process can be readily followed, the black epiblast cells seeming to spread over and gradually almost enclose the white yolk-cells.

This apparent spreading of the epiblast does not take place equally fast all round its margin, and at one place the epiblast, instead of extending over the yolk-cells, bends inwards towards the interior of the egg. This place is visible externally as a sharply defined horizontal or slightly crescentic groove, concave downwards, bounded above by the small black epiblast cells, and below by the large white yolk-

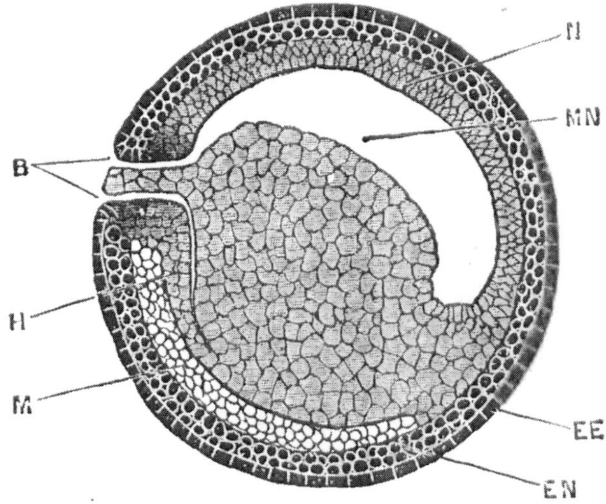

FIG. 25.—Longitudinal vertical section through a frog embryo showing the completion of the mesenteron.

B, blastopore ; EE, epidermic layer of epiblast ; EN, nervous layer of epiblast ; H, invaginate hypoblast ; M, mesoblast ; MN, mesenteron ; N, notochord.

cells. As the epiblast continues spreading over the rest of the yolk this groove becomes horseshoe-shaped, and a little later circular.

If we examine the egg over a mirror, it will be seen that this groove does not remain stationary but shifts gradually from near the equator (DL, Fig. 22A) towards the lower pole, and when it has reached this point it has assumed a circular form (Fig. 24A). In other words the dorsal, lateral, and ventral lips of the blastopore appear successively during its downward movement. From this point its progress is again shifted. The egg rotates about its horizontal axis through about 100° owing to the development within it of an eccentric cavity which causes it to topple over. This brings the original vertical axis of the egg downwards and the blastopore up to the equator again.

The egg has now the appearance shown in Fig. 24, the epiblast covering the whole surface except a circular patch, where alone the yolk-cells are visible from the surface. This circular aperture in the epiblast is called the **blastopore**; it is situated at what will become the posterior end of the embryo; and it is bordered by a distinct rim or lip, round which the epiblast turns inwards into the interior of the egg. The circular plug of yolk-cells filling up the blastopore is spoken of as the **yolk plug.**

The structure of the egg, or rather of the embryo at a slightly later stage, is shown in Fig. 25, which represents a vertical section passing through the middle of the blastopore.

The epiblast covers the whole surface except at the blastopore. From the lip of the blastopore a layer of cells appears to grow into the egg concentrically with its surface.

This layer is called the **hypoblast**: it appears to be formed partly by differentiation of the subequatorial intermediate cells during overgrowth of the lip of the blastopore, partly by differentiation of yolk-cells; it extends much more rapidly dorsally than ventrally, so that while on the upper or dorsal surface it extends to the anterior end of the embryo, at the sides and below it only extends a very short way as yet.

Between this ingrowing layer of hypoblast and the yolk-cells there is a space. This is a very narrow chink near the blastopore, but farther forwards it dilates on the dorsal surface to form a cavity of some size (Fig. 25, MN), wider from side to side than it is dorso-ventrally. (*Cf.* Fig. 26).

This cavity, which is named the **mesenteron**, is the future alimentary canal: it communicates with the exterior through the blastopore, though the aperture is reduced to a narrow chink and is almost stopped up by the yolk plug. The permanent mouth and anus are not yet formed.

During the process of formation of the mesenteron, the segmentation cavity gets pushed out of place and becomes reduced to a chink between the epiblast and hypoblast. (Fig. 25A, SC.)

The Mesoblast. Between the epiblast and hypoblast a third or intermediate layer of cells, the **mesoblast**, is soon established.

It is a separate layer formed by differentiation in part of the subequatorial intermediate cells, and in part from the yolk-cells lying immediately beneath the epiblast, but quite distinct from it. It extends all round the embryo except along the mid-dorsal line, where the space between the epiblast and hypoblast is occupied by the notochord. It is, for a time, incomplete in front, opposite the segmentation cavity, but soon grows in from the sides so as to fill up the deficiency.

The cells of the mesoblast become early arranged in two parallel layers or sheets, which separate slightly from each other, so as to leave between them a narrow space, which later on becomes the **body cavity** or **cœlom.** (*Cf.* Fig. 26.) In many specimens the mesoblast cells are from the first arranged in two layers.

The Notochord. Along the roof of the mesenteron there is a dorsal cell-mass where the roof of the mesenteron and the lateral sheets of mesoblast are folded together. This is the **notochord**, and is derived from both the intermediate and yolk-cells. (Figs. 24 to 26, N.) It is formed at a very early stage, and serves to slightly stiffen the back of the embryo, and is for some time the only skeleton which it possesses.

It very early splits off from the roof of the mesenteron, except at its hinder end, where it remains for some time in continuity with both hypoblast and epiblast at the lip of the blastopore.

Growth of the hypoblast. The hypoblast is formed in the first instance from the inner sheet of intermediate cells at the lip of the blastopore: its aftergrowth is effected, however, mainly at the expense of the yolk-cells, with which it is in contact. By growth at its margin it gradually creeps round until it extends all round the embryo, the yolk-cells forming part of its ventral wall.

Fate of the germinal layers. From one or other of the three germinal layers—**epiblast, mesoblast,** and **hypoblast**—all parts of the embryo are formed.

The **epiblast,** or outer layer, gives rise to the epidermis covering the body generally, and to the various glandular and other structures derived from the epidermis; to the nervous system, both central and peripheral; to the olfactory and auditory epithelium, to the retina and lens of the eye, and to the other sensory organs; to the epithelial lining of the mouth and anus (stomodæum and proctodæum); and to the pineal and pituitary bodies.

The **hypoblast,** or inner layer, gives rise to the epithelium lining the alimentary canal and its various diverticula, including the glands of the œsophagus, stomach, and intestine, the lungs, the bladder, the bile ducts, gall-bladder, pancreatic ducts, and the hepatic cells of the liver and the secreting cells of the pancreas; the notochord is also formed from hypoblast.

From the **mesoblast,** or middle layer, are derived all structures between the epiblast and hypoblast; *i.e.*, the connective tissue, muscles, skeleton (except the notochord), bloodvessels and lymphatics; and also the peritoneum, and the urinary and reproductive organs.

F. Development of the Nervous System.

It is convenient from the point we have now reached to deal with the several systems one by one. The nervous system is a suitable one to commence with, as it appears at a very early stage of development, and plays an important part, especially in the younger stages, in determining the shape and proportions of the embryo.

The epiblast consists almost from the first of two layers, the distinction between which is already established at the close of segmentation. (Fig. 23.) Of these the upper or **epidermic layer** is a single stratum of closely fitting cubical cells; while the lower or **nervous layer** consists of ovoid or spherical cells, more loosely compacted, and two or three deep. It is from the latter that the nervous system is developed.

The first trace of the nervous system is seen about a week after fertilisation, when the embryo is still spherical and the blastopore has become much reduced in size and difficult to see. (*Cf.* Fig. 26.)

The dorsal surface of the embryo now flattens slightly, and along the flattened area the nervous layer of the epiblast thickens to form the **neural plate**, which is wide in front but

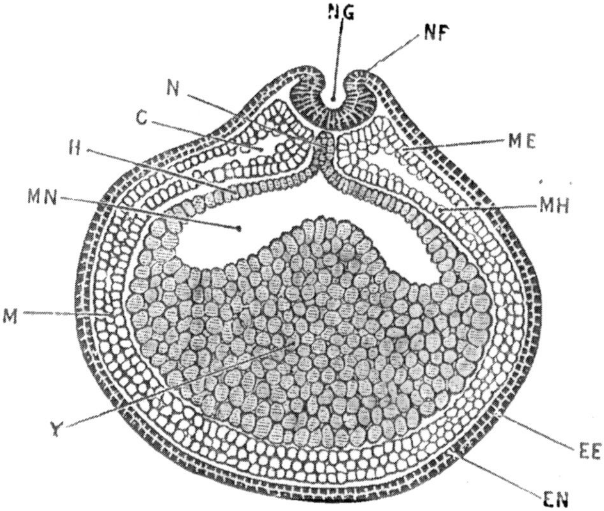

FIG. 26.—Transverse section through a frog embryo during the formation of the neural canal.

C, cœlom ; EE, epidermic layer of epiblast ; EN, nervous layer of epiblast ; H, hypoblast ; M, mesoblast ; ME, somatopleuric layer of mesoblast ; MH, splanchnopleuric layer of mesoblast ; MN, mesenteron ; N, notochord ; NF, neural fold ; NG, neural groove ; Y, yolk cells.

narrows posteriorly towards the blastopore. Slightly raised ridges, the **neural folds**, soon appear, bordering the sides of the neural plate ; and a longitudinal **neural groove** is formed along its dorsal surface in the median line, extending forwards from the blastopore.

A transverse fold connects the anterior ends of the neural folds together, slightly raising up the anterior end of the neural plate. The neural folds now grow rapidly : the groove between them deepens, and the folds becoming more and more prominent bend in towards each other (Fig. 26) and finally meet and fuse, thereby converting the neural groove into a tube.

The neural folds first meet about the junction of the head and trunk of the future tadpole, from which point the fusion extends rapidly in both directions, forwards and backwards. The last point at which fusion occurs is a little distance behind the anterior end of the tube, at the place where the pineal body will appear later.

In front, the neural tube ends blindly; at its posterior end

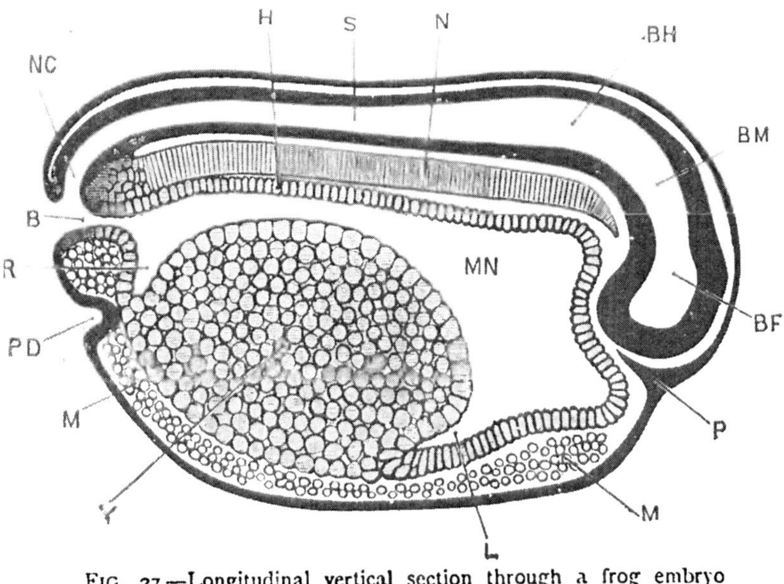

FIG. 27.—Longitudinal vertical section through a frog embryo shortly before the closure of the blastopore. Length of the embryo 2·5 mm. × 30.

B, blastopore; BF, fore-brain; BH, hind-brain; BM, mid-brain; H, hypoblast; L, liver; M, mesoblast; MN, mesenteron; N, notochord; NC, neurenteric canal; P, ingrowth of epiblast to form pituitary body; PD, proctodæum; R, rectal diverticulum of mesenteron; S, central canal of spinal cord; Y, yolk-cells.

it opens to the exterior at the blastopore, and is in free communication with the mesenteron. (*Cf.* Fig. 27.) The short channel of communication between the neural tube and the mesenteron, *i.e.*, between the nervous system and the alimentary canal, is spoken of as the **neurenteric canal**; it is only present for a short time, and closes up before the tadpole hatches.

The neural tube, formed in this way, soon separates from the surface epiblast, and by thickening of its walls and other changes becomes converted into the central nervous system; the anterior part forming the brain, and the posterior part the

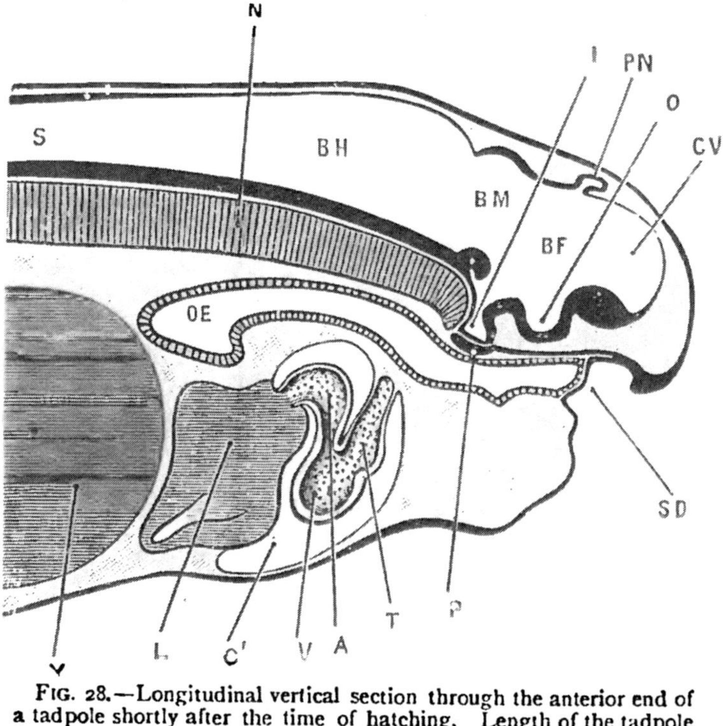

FIG. 28.—Longitudinal vertical section through the anterior end of a tadpole shortly after the time of hatching. Length of the tadpole 8 mm.

A, auricle of heart; BF, fore-brain; BH, hind-brain; BM, mid-brain; C', pericardial cavity; CV, vesicle of cerebral hemispheres; I, infundibulum; L, liver; N, notochord; O, depression of floor of fore-brain from which the optic vesicles arise; OE, œsophagus; P, pituitary body; PN, pineal body; S, central canal of spinal cord: SD, stomodæum; T, truncus arteriosus; V, ventricle; Y, yolk-cells.

spinal cord. The lumen or cavity of the tube persists as the central canal of the spinal cord and the ventricles of the brain.

The Brain. At the time of its first appearance the brain is bent at right angles about the middle of its length; the axis of the anterior portion being vertical, and that of the posterior portion horizontal. (Fig. 27.) The posterior portion, or **hind-brain**, **BH**, is wide from side to side, and has moderately thick sides and floor, but a thin roof; it is continuous behind with the spinal cord.

The anterior or vertical portion has walls of nearly uniform thickness in all parts. It is divided by a slight constriction,

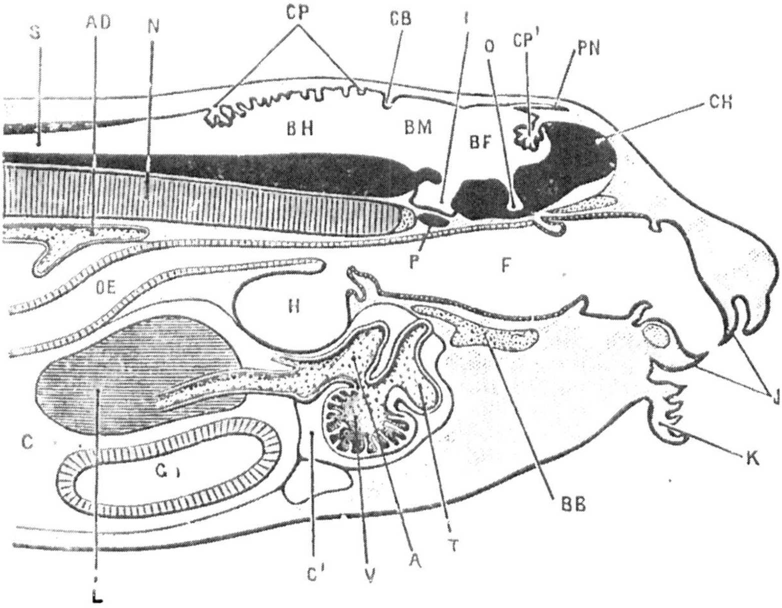

FIG. 29.—Longitudinal vertical section through the head and anterior part of the body of a tadpole about the time of appearance of the hind legs. Length of tadpole, 12 mm. × 14.

A, auricle of heart; AD, dorsal aorta; BB, basi-branchial cartilage; BF, fore-brain; BH, hind-brain; BM, mid-brain; C, cœlom or body cavity; C', pericardial cavity; CH, cerebral hemisphere; CB, rudimentary cerebellum; CP, choroid plexus of fourth ventricle; CP', choroid plexus of third ventricle; F, pharynx; G, stomach; H, lung; I, infundibulum; J, horny jaws; K, lip; L, liver; N, notochord; O, depression of floor of fore-brain from which the optic nerves arise; OE, œsophagus; P, pituitary body; PN, pineal body; S, central canal of spinal cord; T, truncus arteriosus; V, ventricle.

best marked at the sides, into an upper or posterior part, the **mid-brain**, **BM**, which forms the angle of the bend and lies opposite the anterior end of the notochord; and a lower and larger portion, the **fore-brain, BF**, which is produced laterally into a pair of hollow outgrowths, the **optic vesicles**.

The further development of the brain is illustrated by Figs. 28 and 29. It will be seen that the rectangular bending of the brain, which is known as the **cranial flexure**, and which was so prominent a feature in the earlier stage, is no longer obvious; a closer comparison of the figures will show, however, that this straightening of the brain, or rectification of the cranial flexure, is apparent rather than real, and is brought about partly by the development of the cerebral hemispheres, which grow upwards and forwards from the fore-brain, and still more largely by the formation of the mouth and the growth forwards of the face and lips, which cause the brain to take a much less prominent share in determining the shape of the head.

The **hind-brain, BH**, has undergone but little change in Fig. 28, except an increase in thickness of its floor and sides. At the stage represented in Fig. 29 it is separated from the mid-brain on the dorsal surface by a well-marked groove, immediately behind which the roof of the hind-brain is thickened transversely to form the **cerebellum, CB**. The cavity of the hind-brain remains as the **fourth ventricle**, the roof of which is very thin and thrown into numerous transverse folds, **CP**, which hang down into the ventricle, and between the layers of which lie the bloodvessels of the choroid plexus of the ventricle.

The **mid-brain, BM**, thickens on its floor to form the **crura cerebri**. Its roof grows out laterally into a pair of hollow ovoid processes, the **optic lobes**; and its cavity persists as the **aqueductus Sylvii**, or iter a tertio ad quartum ventriculum.

The **fore-brain, BF**, becomes the **thalamencephalon** of the adult; its cavity becomes the **third ventricle**, which by thickening of its walls to form the **optic thalami** is reduced to a vertical cleft, very narrow from side to side. Its floor is produced downwards and backwards into a hollow sac-like diverticulum, the **infundibulum, I**, in connection with which is the pituitary body. In front of the infundibulum is a transverse ridge projecting into the ventricle, and formed by the roots of the optic nerves.

The roof of the fore-brain remains thin; a little behind the middle of its length the **pineal body, PN**, arises as a median hollow diverticulum, Figs. 28 and 29; this is formed at the spot where the final closure of the neural tube took place, and is at first directed backwards; in the later stages it grows forwards and forms a rounded vesicle connected with the brain by a long pigmented stalk; when the skull develops it cuts off the vesicle from the stalk, the former remaining as a small rounded body outside the skull, while the stalk persists as a slender pigmented tract within the cranial cavity.

In front of the pineal body, and at the anterior end of the fore-brain, the roof is thrown into folds which hang down into the ventricle forming a choroid plexus, **CP'**, similar to that in the medulla.

The anterior end of the fore-brain grows forwards as a median thin-walled **cerebral vesicle**, from which at a slightly later stage the **cerebral hemispheres**, **CH**, arise as a pair of hollow outgrowths; the **foramina of Monro** being the apertures of communication between the **lateral ventricles** or cavities of the hemispheres, and the third ventricle. The anterior ends of the hemispheres grow forwards as the **olfactory lobes**, which become fused together in the median plane.

The **peripheral nervous system**. The cranial nerves and the dorsal roots of the spinal nerves are formed from the deeper or nervous layer of the epidermis. They appear to arise as lateral outgrowths from the edges of the neural plate, and may be recognised at a very early stage, while the neural groove is still shallow and open; they are, therefore, at their first appearance continuous with the brain or spinal cord.

The ventral roots of the spinal nerves arise later than the dorsal ones, as outgrowths from the cord near its ventral surface. They are at first independent of the dorsal roots, but soon become connected with these.

G. Development of the Sense Organs.

The organs of special sensation are developed from the deeper or nervous layer of the epiblast, and become connected with their respective nerves at a very early stage of their formation.

The derivation of the sense organs from the epiblast is explained by the fact that they are concerned with the appreciation of the presence and nature of external objects, and are therefore necessarily formed on the surface of the body. They may be regarded as specially modified portions of the epidermis.

The Nose. The olfactory organs appear at a very early stage as paired thickenings of the nervous layer of the epiblast at the anterior end of the head, in the angles between the fore-brain and the optic vesicles. A pitting-in of the surface, involving both layers of the epiblast, soon appears in each of these thickenings, and the pits so formed become the nasal sacs; the mouths of the pits forming the nostrils or **anterior nares**, and the epiblast lining the pits giving rise to the olfactory epithelium.

From the inner or deeper end of each olfactory pit a diverticulum, at first solid, but soon becoming hollow, grows downwards to the roof of the pharynx, into which it opens, as the **posterior nares**, very shortly after the formation of the mouth opening.

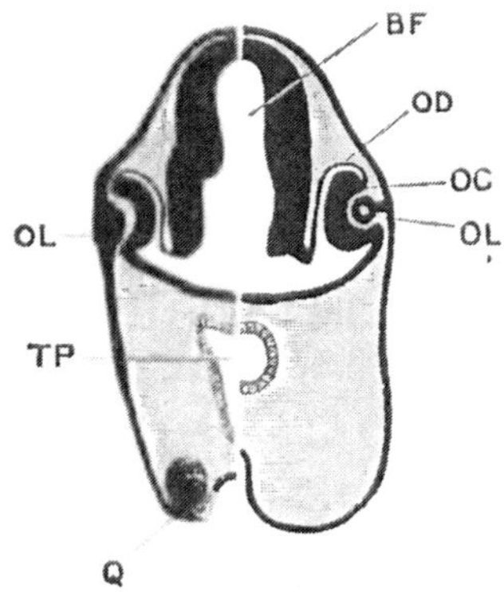

FIG 30.—Half sections in the transverse plane of a tadpole 10 mm. long (left half) and of a tadpole 12 mm. long (right half). × 35.
BF, fore-brain ; OD, outer wall of optic cup (pigment layer of adult retina); OC, inner wall of optic cup (remainder of adult retina); OL, lens, attached to epiblast in younger tadpole, but forming a hollow vesicle at the later stage; TP, pharynx ; Q, sucker. [G. H. F.]

The Eye. The eye differs from the other sense organs, inasmuch as the lens alone is formed directly from the surface epiblast, while the sensitive part of the eye, or retina, arises as an outgrowth from the brain. The optic vesicles have already been described as arising at a very early period as lateral outgrowths from the fore-brain; these soon become constricted at their necks so as to be connected with the brain by narrow stalks, which ultimately become the optic nerves.

The outer surface of each optic vesicle, which is at first in close contact with the surface epiblast, soon becomes flattened (Fig. 30, left half), and then thickens so greatly as almost to obliterate the cavity of the vesicle. At the same time a thickening of the deeper or nervous layer of the surface epiblast takes place opposite the optic vesicle; this grows rapidly and forms a spherical body, projecting inwards from the surface; this is at first solid, but soon becomes hollow and breaks away completely from the surface epiblast; it becomes later on the lens of the eye, and may be spoken of as the **lens vesicle.**

Partly in consequence of the ingrowth of the lens vesicle, and partly through growth of the optic vesicle itself, this latter becomes pitted on its outer surface, and so converted into a cup —the **optic cup**—with double walls; the inner wall being the thickened and originally outer wall of the optic vesicle, and the outer wall of the cup being the original inner or deeper part of the wall of the vesicle. The lip of the cup is incomplete below, owing to the presence of a slit, the **choroidal fissure,** through which mesoblastic elements penetrate into the interior of the eye.

From the optic cup and lens vesicle the adult eye is derived in the following way: The lens becomes solid, owing to thickening of its inner wall, which proceeds so far as to finally obliterate the cavity. The optic cup enlarges considerably; it remains in contact with the lens at its edge or lip, but elsewhere is separated from it by a space which becomes the posterior chamber of the eye, and in which the vitreous humour is formed. The inner wall of the optic cup gives rise to the retina, the rods and cones growing out from its outer surface; while the outer and thinner wall of the optic cup forms the layer of pigment cells in which the rods and cones are imbedded. The choroid and sclerotic coats are formed from the mesoblast surrounding the optic cup.

The eye develops very slowly, and throughout the tadpole stage of existence is in a very rudimentary and imperfect condition.

The Ear. The ears are developed as a pair of pit-like invaginations of the nervous layer of the epiblast at the sides of the hind-brain. The invaginations do not involve the epidermic or surface layer of the epiblast, so that the auditory pits do not open to the exterior.

The mouths of the pits very early narrow and close; and the **auditory vesicles** so formed separate from the epiblast and lie in the mesoblast at the sides of the head. The vesicle becomes the vestibule of the adult ear; the semicircular canals arising as outgrowths from it.

Throughout the tadpole stage of existence there is no further modification; but shortly after the metamorphosis the hyomandibular cleft, which has at no period opened to the exterior, is stated to widen somewhat and form the Eustachian passage, while the layer of integument closing its outer end becomes the tympanic membrane. There is some reason, however, for thinking that the Eustachian passage develops independently in the frog, and not from the hyomandibular cleft. The columella, which has been described with the skull, is formed still later. (*Cf.* Fig. 10, p. 48.)

Special Sense Organs. During the tadpole stage, while the animal is leading an aquatic life, special sense organs in the form of small epidermal papillæ, supplied by branches of the trigeminal and pneumogastric nerves, are found arranged in rows along the body, and round the eyes, and in other parts of the head. These are lost at the time of the metamorphosis.

The mouth of the tadpole is also provided with small rounded papillæ, which are probably organs of taste. (See Fig. 29.)

H. Development of the Alimentary Canal.

The alimentary canal is developed in three lengths: (1) the **mesenteron,** which is formed by splitting apart of the yolk-cells as described above; this gives rise to nearly the whole length of the alimentary canal; and from it are developed the gill slits, the lungs, the thyroid, the liver, the pancreas, and the bladder; as well as the notochord; (2) the **stomodæum,** which is a pitting-in at the anterior end of the body, from which the mouth and pituitary body are formed; and (3) the **proctodæum,** which is a similar pitting-in at the hinder end of the body to form the anal or cloacal opening.

From the mode of their formation it follows that the mesenteron is lined by hypoblast, and the stomodæum and proctodæum by epiblast.

1. The **mesenteron.** The early development of the mesenteron has already been described.

The anterior end of the mesenteron, in the head region, is considerably dilated from the first: and at the hinder end of the embryo a similar, though much smaller, expansion takes place. In this way (*cf.* Fig. 27), the mass of the food-yolk becomes confined to the ventral portion of the body region, not extending into either the head or the tail.

The **hypoblast**, which is a definite layer of cells, at first confined to the roof of the mesenteron, gradually spreads round its sides until it encloses the whole of the food-yolk, and the alimentary canal is completed as a tube, which from the first is slightly convoluted. When the tadpole begins to feed, the alimentary canal lengthens rapidly, and becomes coiled in a spiral manner. Except at the anterior end, in the gill-bearing region, it is of approximately uniform diameter throughout. During the metamorphosis, the alimentary canal shortens considerably, and the distinction between stomach, small intestine, and large intestine, is definitely established.

The **liver** is recognisable at a very early stage (Fig. 27) as a ventral and backwardly directed diverticulum of the anterior part of the mesenteron, forming the anterior boundary of the mass of food-yolk. In the later stages the walls of the diverticulum thicken, and become thrown into folds between which the vascular mesoblast makes its way: the diverticulum itself persists as the bile duct, and the gall bladder arises as an outgrowth from this.

The **pancreas** is developed as a pair of hollow outgrowths from the mesenteron, behind the liver: in the later stages the ducts shift so as to open into the bile duct instead of directly into the intestine.

The **bladder** arises shortly before the metamorphosis as a ventral outgrowth from the hinder end of the mesenteron, which soon becomes bifid at its distal blind end.

The **post-anal gut** is an extension of the hinder end of the mesenteron into the base of the tail, which appears as this latter is developed: it becomes solid after a short time, and later on disappears altogether. It is perhaps to be regarded as formed by a mechanical drawing out of the intestine by the outgrowing tail.

The **lungs**. Immediately behind the gill-bearing region or pharynx, the alimentary canal narrows very considerably; its sides become folded inwards, and the two folds meeting each other divide the canal into a dorsal tube or œsophagus, and a ventral one which forms the laryngeal chamber: from this latter the lungs arise as thin-walled lateral outgrowths. They appear first in young tadpoles of about 8 mm. length, *i.e.*, some time after hatching, but shortly before the opening of the mouth. About the time that the lungs first appear, in tadpoles of about 8 mm. length, the œsophagus, which up to this time has been tubular, becomes solid, and remains so until a short time after the formation of the mouth. The meaning of this curious point has not been ascertained.

2. The **stomodæum**. At the stage represented in Fig. 27, shortly after closure of the neural canal, a conical ingrowth, P, of the nervous layer of the epidermis is formed at the anterior end of the body immediately below the fore-brain: from this ingrowth the pituitary body is developed, and a slight depression of the surface epiblast opposite its base, marks the position of the stomodæum.

At the time of hatching, this depression is a small shallow pit, separated from the anterior end of the mesenteron by a thin septum. Soon after hatching, in tadpoles of about 9 mm. length, this septum becomes perforated, and the alimentary canal communicates with the exterior through the stomodæal pit. After the perforation is effected, the lips with the whole anterior part of the face grow forwards rapidly; the horny jaws are formed, and the tadpole begins to feed vigorously. (*Cf.* Figs. 28 and 29.)

The **pituitary body** (Figs. 27 to 29, P) is formed from the ingrowing stalk of epiblast described above: this rapidly elongates, growing backwards between the brain and the roof of the mesenteron until it reaches the infundibulum; its hinder end now becomes tubular, gives off a few lateral diverticula; separates from the stalk, which soon disappears, and becomes applied to the ventral surface of the hinder end of the infundibulum to form the pituitary body.

3. The **proctodæum** or anal invagination appears before the stomodæum. Shortly before the neural folds have met to form the neural tube, the proctodæum is visible as a small median depression of the epiblast at the hinder end of the embryo, a little way below the blastopore. The cells lining it are rather strongly pigmented, and slightly larger than the surrounding epiblast cells.

From the hinder end of the mesenteron a rectal diverticulum (Fig. 27, R) extends downwards towards the proctodæum; a little later, and some time before the tadpole hatches, the two structures meet; perforation occurs; and the definitive anal or

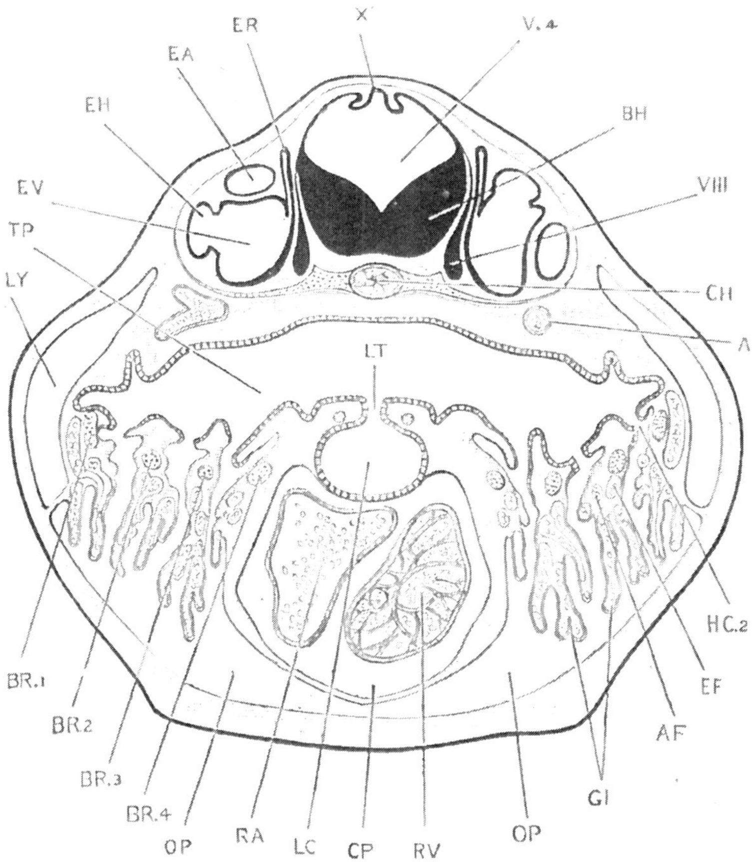

FIG. 30A.—Transverse section through the head of a 12 mm. tadpole; the section passing through the auditory organs, the pharynx and opercular gills, the glottis and laryngeal chamber, and the heart. × 40. A, aorta; AF, afferent blood-vessel of second branchial arch; BH, hind brain; BR.1, .2, .3, .4, first, second, third, and fourth branchial arches; CH, notochord; CP, pericardial cavity; EA, anterior vertical semicircular canal; EF, efferent blood-vessel of second branchial arch; EH, horizontal, semicircular canal; ER, recessus vestibuli; EV, vestibule of ear; GI, opercular gills; HC.2, second branchial cleft; LC, laryngeal chamber; LT, glottis; LY, lymphatic space; OP, opercular cavity; RA, auricle of heart; RV, ventricle; TP, pharynx; V.4, fourth ventricle; X', choroid plexus of fourth ventricle; VIII, auditory nerve.

To face p. 129.

cloacal opening is formed. For a short time the blastopore and the proctodæum are both open; but very shortly after completion of the proctodæum the blastopore closes finally.

I. The Gill Clefts and Arches.

Some little time before the tadpole is hatched a series of vertical ridge-like thickenings appear on the sides of the head and neck. These are the **visceral arches,** and are six in number on each side.

The most anterior is the **mandibular arch,** and gives rise later on to the lower jaw; the second is the **hyoid arch**; and the succeeding four are the first, second, third, and fourth **branchial arches** respectively.

About the time of hatching the **external gills** grow out as branching and richly ciliated processes from the outer surfaces of the first and second branchial arches, and a little later from the third branchial arches as well (Fig. 31).

At the same time, the hypoblastic epithelium at each side of the buccal cavity becomes thrown into folds, which extend outwards towards the surface of the neck as paired outgrowths, lying between the visceral arches. Of these outgrowths or pouches, which are known as **visceral clefts,** there are five on each side. The most anterior one is the **hyomandibular cleft,** and lies between the mandibular and hyoid arches: its outer end lies very close to the surface of the neck, though it does not actually open to the exterior.

The four hinder visceral clefts perforate the skin about the time of formation of the mouth opening, *i.e.*, in tadpoles of about 9 mm. length, and open to the exterior as the **gill clefts.** These are slit-like openings lying between the hyoid and first branchial, first and second branchial, second and third branchial, and third and fourth branchial arches respectively; and are known as the first, second, third, and fourth branchial clefts.

From the hyoid arches a pair of opercular folds arise, which grow back over the external gills, and the branchial arches and clefts. The two opercular folds meet below the neck in the mid-ventral line, and enclose the gills in a **branchial chamber.** The hinder borders of the opercular folds fuse with the body-wall except at one place on the left side, where a spout-like opening remains, through which the branchial chamber communicates with the exterior.

As the opercular folds develop, the first formed external gills gradually shrivel up, and are replaced by a second set enveloped by the ectoderm on the lower and outer side of the arches. These latter are delicate thin-walled vascular tufts, arranged in a double row along the ventral half of each of the first three branchial arches, and in a single row on the fourth branchial arch.

The inner borders of the branchial arches are thickened, and produced into processes which unite to form a kind of filtering apparatus, or sieve, through which the water, taken in through the mouth or nose, is strained before being passed over the gills into the branchial cavity and so out.

K. The Vascular System.

The heart is at first a straight tube developed in the mesoblast of the ventral wall of the pharynx. This soon lengthens, becomes twisted into an S shape, and divided by transverse constrictions into chambers. (*Cf.* Figs. 28, 29, and 32.) The auricle is at first single, but later becomes divided by the downgrowth of a septum from its dorsal wall.

While the tadpole is breathing by means of gills, its circulation is in all essential respects that of a fish. The venous blood, returned from the body generally, enters the posterior end of the heart, or sinus venosus: from this it passes into the second or auricular chamber, thence to the ventricle, and from that to the truncus arteriosus. From this latter arise on each side the aortic arches, which carry the venous blood to the gills to be aërated: from the gills the blood is collected by efferent vessels, which unite above the alimentary canal to form the dorsal aorta, which by its branches distributes the arterialised blood to all parts of the body.

1. The Circulation during the time the tadpole is breathing by its external gills.

The arrangement of the bloodvessels, and the course of the circulation in a $6\frac{1}{2}$ mm. tadpole, at a time when the external gills are in full activity, is shown in Figs. 31 and 32.

The truncus arteriosus, on reaching the anterior end of the pericardial cavity, divides at once into right and left branches. Each of these again divides into two, the afferent vessels for the first and second branchial arches, AF_1 and AF_2, which carry blood into the external gills and their branches: from these the blood passes through short wide capillary loops into the efferent branchial vessels, EF_1 and EF_2, which carry it, now aërated, to

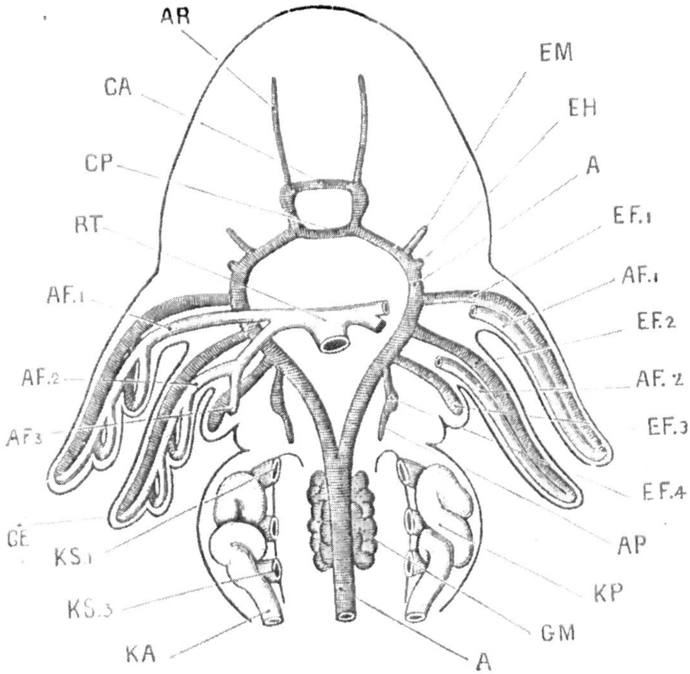

FIG. 31.—Diagrammatic figure of the head and fore part of the body of a 6½ mm. tadpole, showing the arrangement of the branchial vessels as seen from the ventral surface. The heart has been removed. × 33.

A, dorsal aorta; AF_1, AF_2, AF_3, afferent branchial vessels of the first, second, and third branchial arches; AP, pulmonary artery; AR, anterior cerebral artery; CA, anterior commissural artery; CP, posterior commissural artery: the arterial circle formed by these commissural vessels with the carotid arteries surrounds the infundibulum of the brain: EF_1, EF_2, EF_3, EF_4, efferent branchial vessels of the first, second, third, and fourth branchial arches; EH, efferent hyoidean vessel; EM, efferent mandibular vessel; GE, external gill; GM, glomerulus; KA, segmental or archinephric duct; KP, head kidney or pronephros; KS_1, KS_3, first and third nephrostomes of pronephros; RT, truncus arteriosus.

the dorsal aorta in the roof of the pharynx. The dorsal aortæ of the two sides run forwards as the carotid arteries, AC, to supply the head and brain, and also run backwards in the roof of the pharynx, the aortæ of the two sides meeting and uniting about the junction of head and body to form the single systemic aorta which supplies arterial blood to all parts of the body.

Besides the complete sets of afferent and efferent branchial vessels in the first and second branchial arches, similar vessels,

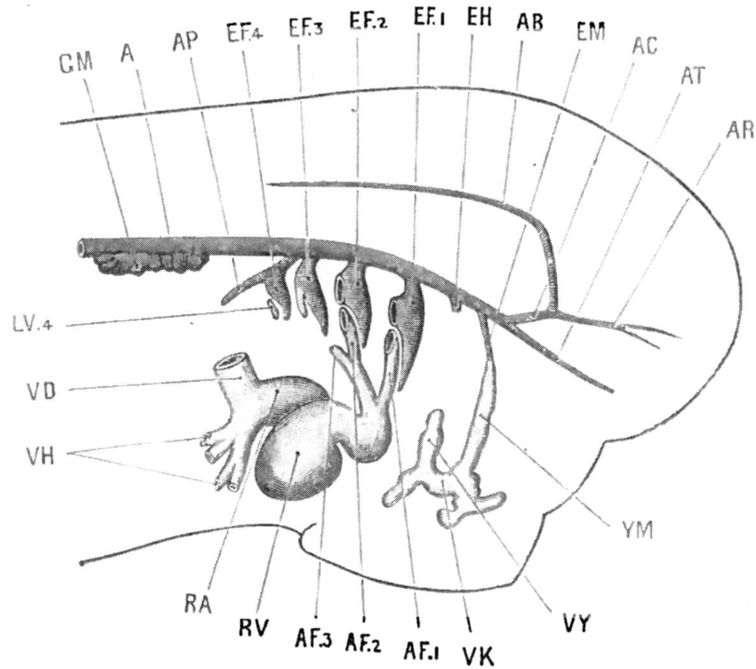

FIG. 32.—Diagrammatic figure of the head and fore-part of the body of a 6½ mm. tadpole, showing the heart, aorta, and vessels of the branchial arches from the right side. The external gills have been removed. × 40.

A, dorsal aorta; AB, basilar artery; AC, carotid artery; AF_1, AF_2, AF_3, afferent branchial vessels of first, second, and third branchial arches; AP, pulmonary artery; AR, anterior cerebral artery; AT, anterior palatine artery; EF_1, EF_2, EF_3, EF_4, efferent branchial vessels of first, second, third, and fourth branchial arches; EH, efferent hyoidean vessel; EM, efferent mandibular vessel; GM, glomerulus; LV_4, lacunar afferent vessel of fourth branchial arch; RA, auricle; RV, ventricle; VD, Cuvierian vein; VH, hepatic veins; VK, vein of sucker; VY, hyoidean vein; YM, mandibular vein.

as yet incompletely developed, are present in the hinder arches as well.

In the third branchial arch, there is a short afferent branch,

AF_3, from the afferent vessel of the second branchial arch, which as yet ends blindly. There is also a well-developed efferent vessel, EF_3, which opens into the dorsal aorta.

In the fourth branchial arch there is no afferent vessel, but an efferent vessel, EF_4, is present, opening into the dorsal aorta. From this efferent vessel, just before it reaches the aorta, a backwardly directed branch arises, which will become later the pulmonary artery, AP.

In front of the first branchial arch, vessels are present in the hyoid and mandibular arches, which clearly belong to the same category as the branchial vessels, but which never attain full development, probably owing to the fact that no gills are formed on these arches. Efferent branches, EH, EM, opening into the dorsal aorta, are present in both hyoid and mandibular arches; but these have no connection with the heart, as there are no afferent vessels corresponding to them.

The condition of the bloodvessels, while the tadpole is breathing by external gills, may be summarised thus:—Complete systems of afferent and efferent vessels, connecting the heart with the aorta through the gill capillaries, are present in the first and second branchial arches, and at a stage slightly later than that shown in Fig. 32 in the third branchial arch as well. A similar set of vessels, but incomplete, is present in the fourth branchial arch: and vessels formed on the same plan, but still less complete, and showing signs of degenerative changes, are present in the hyoid and mandibular arches.

There are thus six sets of branchial vessels on each side of the pharynx: of these, three, in the first, second, and third branchial arches, are complete; one, in the fourth branchial arch, is incomplete; and two, in the hyoid and mandibular arches, are rudimentary.

2. The Circulation during the time the tadpole is breathing by its opercular gills.

On the formation of the gill-slits, additional loops of communication are formed in the gill tufts between the afferent and efferent vessels of the first, second, and third branchial arches, and also a series of similar loops between the afferent and efferent vessels of the fourth branchial arch. The vessels in the hyoid and mandibular arches undergo further retrograde changes, and need not be described in detail.

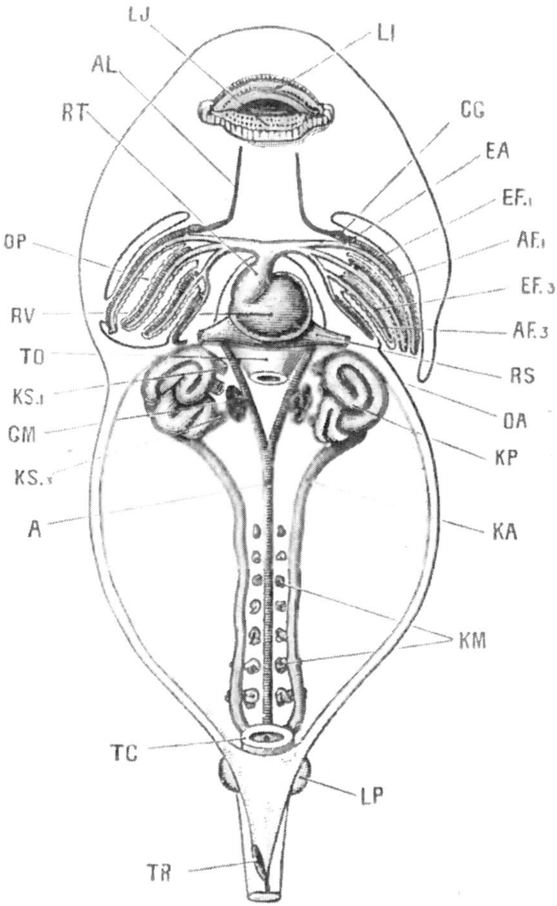

FIG. 33.—A 12 mm. tadpole dissected from the ventral surface to show the heart, the internal gills, the branchial vessels, and the head kidneys and their ducts. The tail, which is about double the length of the head and body, has been removed. × 22.

A, dorsal aorta; AF_1, AF_3, afferent branchial vessels of first and third branchial arches; AL, lingual artery; CG, carotid gland; EA, junction between afferent and efferent branchial vessels of first branchial arch; EF_1, EF_3, efferent branchial vessels of first and third branchial arches; GM, glomerulus; KA, archinephric or segmental duct; KM, Wolffian tubules; KP, pronephros or head kidney; KS_1, KS_3, first and third nephrostomes of head kidney; LI, upper lip; LJ, lower lip; LP, hind limb; OA, aperture of opercular cavity; OP, opercular cavity; RS, sinus venosus; RT, truncus arteriosus; RV, ventricle TC, cloaca; TO, œsophagus, cut short; TR rectal spout.

In tadpoles of 12 mm. length, in which the opercular gills are fully established, and the external gills shrivelling up, the condition of the bloodvessels is shown in Figs. 33 and 34.

The truncus arteriosus divides at once into right and left branches, which run straight outwards in the floor of the

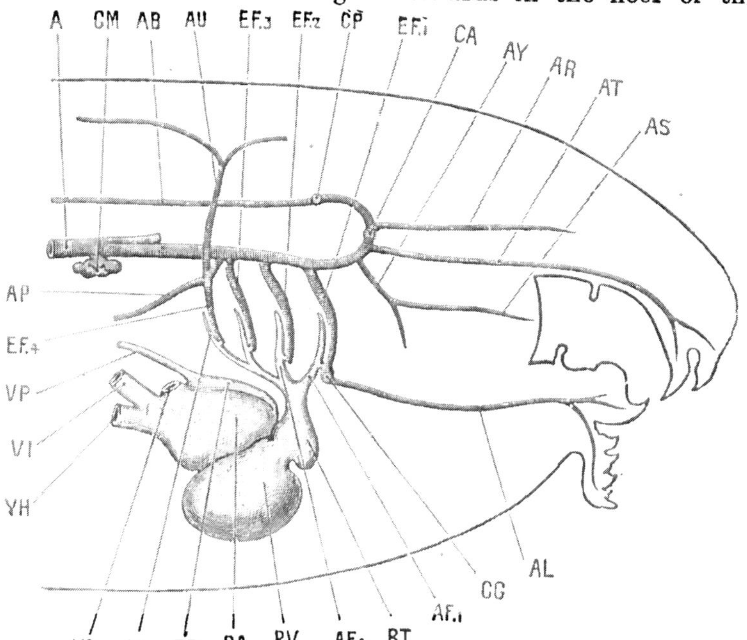

FIG. 34.—A diagrammatic figure of the head and neck of a 12 mm. tadpole from the right side to show the heart and branchial vessels. The gills and the gill capillaries are not represented. × 35.

A, dorsal aorta; AB, basilar artery; AF_1, AF_2, AF_4, afferent branchial vessels of first, second, and fourth branchial arches; AL, lingual artery; AP, pulmonary artery; AR, anterior cerebral artery; AS, posterior palatine artery; AT, anterior palatine artery; AU, cutaneous artery; AY, pharyngeal artery; CA, anterior commissural vessel; CG, carotid gland; CP, posterior commissural vessel; EF_1, EF_2, EF_3, EF_4, efferent branchial vessels of first, second, third, and fourth branchial arches; GM, glomerulus; RA, right auricle; RB, left auricle; RT, truncus arteriosus; RV, ventricle; VD, Cuvierian vein; VH, hepatic vein; VI, posterior vena cava; VP, pulmonary vein.

pharynx. Each of these branches divides, after a short course, into three vessels, and the hindmost vessel again into two. In this way the four afferent branchial vessels, AF_1, AF_2, AF_3, AF_4, of the first, second, third, and fourth branchial arches respectively are formed.

Each afferent vessel runs outwards and upwards in its own arch. The efferent branchial vessels lie immediately in front of the corresponding afferent vessels, with which they are connected by very numerous capillary loops in the substance of the internal gills, and not shown in the figures. At their upper ends the efferent vessels open, as before, into the dorsal aorta, Fig. 34.

The venous blood in the heart is driven by the contraction of the ventricle into the truncus arteriosus, and then along the afferent branchial vessels, through the capillary loops of the gills, in which it gets aërated, to the efferent branchial vessels; and thence to the dorsal aorta, and so all over the body.

The lungs are by this time of considerable size: they receive blood by the pulmonary arteries, AP, which, as already noticed, are branches from the efferent vessels of the fourth branchial arches, and therefore contain blood which has already passed through the gill capillaries. The blood from the lungs is returned direct to the heart by two pulmonary veins which unite and open into the left auricle, the single auricular cavity of the earlier stage being by this time divided by a vertical septum into right and left auricles.

One other point of great importance remains to be noticed in the arrangement of the branchial vessels of the tadpole. The afferent and efferent vessels of each arch at first communicate only through the gill capillaries: but in tadpoles of about 12 mm. length each efferent vessel becomes directly connected at its ventral end with the corresponding afferent vessel, Fig. 34. These direct connections are situated ventrally to the gills, so that the blood in any one of the afferent branchial vessels has two paths open to it: it may either (1) continue along the afferent vessel, and then reach the efferent vessel by passing through the connecting loops afforded by the gill capillaries; or (2) it may pass at once into the efferent vessel through the direct communication, and so reach the dorsal aorta without having passed through the gill at all.

So long as the tadpole is breathing by gills, these direct communications between afferent and efferent vessels, though present in all four branchial arches, are so small that practically no blood passes through them, and all the blood is compelled to pass through the gills to reach the aorta.

3. **The Changes in the Circulation at the time of the Metamorphosis.**

At the time of the metamorphosis, however, when the anterior limbs are protruded, and the tail begins to shorten, these direct communications enlarge, so that an increasing amount of blood takes the direct short passage, and reaches the aorta without having passed through the gills. Additional work is thus thrown on the lungs and skin, which consequently receive a larger supply of blood: the gills rapidly atrophy, though remnants of them usually persist, in a functionless condition, until the end of the first year; and the change from the gill-breathing to the air-breathing condition is completed.

The further changes necessary to convert the circulation into that of the adult are slight. Of the four aortic arches present at the metamorphosis (Fig. 34), the first, in the first branchial arch, persists as the carotid arch of the adult frog; the lingual artery is a branch from the ventral end of the efferent vessel of the arch, and is present from an early stage of development (Fig. 33); and the external and internal carotid arteries are already present. The carotid gland, CG, is not, as sometimes stated, a persistent portion of a gill, but is formed by further elaboration of the direct communication between the afferent and efferent branchial vessels of the first branchial arch.

The second aortic arch, in the second branchial arch, becomes the systemic arch of the frog. Its dorsal end remains connected with the carotid arch, though the connection may in the adult become closed and ligamentous. (*Cf.* Fig. 5, p. 29.)

The third aortic arch, in the third branchial arch, loses its connection with the aorta, and finally disappears altogether.*

The fourth aortic arch, in the fourth branchial arch, also

* According to another view, of fairly general acceptance the following is the scheme of arterial arches in the frog:

Arch.		Embryonic.		Adult.
I.	...	Mandibular	...	?
II.	...	Hyoidean	...	?
III.	...	First branchial	...	Carotid arch.
IV.	...	Second branchial	...	Systemic arch.
V.	...	Third branchial	...	Cutaneous artery.
VI.	...	Fourth branchial	...	Pulmonary artery.

loses its connection with the aorta, but persists as the pulmo-cutaneous arch of the adult, from which both pulmonary and cutaneous arteries arise.*

L. Development of the Muscular System and the Cœlom.

The splitting of the mesoblast into outer or **somatopleuric**, and inner or **splanchnopleuric** layers has already been described. (*Cf.* Fig. 26, p. 118.)

In the body the mesoblast becomes very early divided on each side into (1) a **vertebral plate**, which is dorsally situated, and lies alongside of the spinal cord and notochord; and (2) a **lateral plate**, which surrounds the side of the body.

The vertebral plate very early becomes divided transversely into muscle-segments or **myotomes**, which form a row of hollow and somewhat cubical bodies, lying along each side of the spinal cord, and separated from each other by connective tissue septa. Later on, the walls of the myotomes thicken considerably, especially the inner walls, and become converted very largely into muscles; while the cavities become obliterated.

The myotomes may be well seen in the tail of the tadpole, where they form the great lateral sheets of muscle on each side of the tail, by which the swimming movements are effected. Owing to the transparency of the tail, their arrangement can be very readily made out; the septa dividing the successive myotomes from each other are not transverse, but > shaped, with the angles directed forward towards the head.

The lateral plates are also in part converted into muscle; the two layers, somatopleuric and splanchnopleuric, remain comparatively thin, but the space between them widens out considerably, and becomes the body cavity or cœlom. This at first consists of two separate halves, right and left; but, owing to the splitting of the mesoblast extending down to the mid-ventral line, the cavities of the two sides soon became continuous. The anterior portion of the cœlom is very early shut off from the hinder part as the pericardial cavity. (*Cf.* Figs. 28 and 29.)

The outer or somatopleuric layer of mesoblast, with the epiblast, forms the body-wall of the adult; the inner or splanchnopleuric layer, with the hypoblast, forms the wall of the alimentary canal and its diverticula. The cells covering the

* See note, p. 137.

free surfaces of both layers, *i.e.*, the cells lining the body cavity, become the peritoneum, or cœlomic epithelium, from which, as we have already seen, the ovaries and testes are formed.

M. Development of the Skeleton.

1. The Vertebral Column.

The earliest skeletal structure, and for a time the only one, is the notochord, the development of which from the hypoblast of the mid-dorsal wall of the mesenteron has already been described. It forms a cellular rod extending from the blastopore to the pituitary body; and as the tail is formed, it extends back into it. The notochord consists of vacuolated cells, filled with fluid, and is invested by a delicate structureless sheath.

About the time of appearance of the hind legs, a delicate skeletal tube, at first soft but soon becoming cartilaginous, is formed round the notochord from the mesoblast. This tube grows upwards at the sides of the spinal cord, as a pair of longitudinal ridges, with which a series of cartilaginous arches, which appeared at the sides of the spinal cord at a slightly earlier stage, very soon become continuous.

By the appearance of transverse lines of demarcation, the cartilaginous sheath of the notochord becomes cut up into a series of nine vertebræ, followed by a posterior unsegmented portion, which later becomes the urostyle. This transverse division does not affect the notochord, which remains as a continuous structure until the complete absorption of the tail at the end of the metamorphosis.

Shortly after the metamorphosis thin rings of bone, slightly constricted in their centres, so as to be hourglass-shaped in section, are developed in the membrane investing the cartilaginous sheath of the notochord: these correspond with the nine vertebræ already present, and form the first rudiments of the vertebral centra.

In the intervertebral regions, between the successive bony rings, annular thickenings of the cartilaginous sheath occur, which grow inwards so as to constrict and ultimately obliterate the notochord. Each of these intervertebral rings becomes, after the metamorphosis, divided into an anterior and a posterior portion, which fuse with the bony centra of adjacent vertebræ, and ossify to form their articular ends.

From the circumference, and from the articular ends of each

vertebra, ossification gradually spreads inwards; but a small portion of notochord persists in the middle of each centrum for a long time, or even throughout life.

The vertebræ are not placed opposite the myotomes, but alternate with these; so that each vertebra is acted on by two myotomes on each side, one pulling it forwards, and the other backwards.

The transverse processes are at first independent of the corresponding vertebræ, but very early fuse with them. They extend into the septa between the myotomes, and probably correspond to the ribs of other vertebrates.

The urostyle is the part of the axial skeleton behind the vertebræ; it is not divided into vertebræ at any stage in development.

The anterior end of the notochord, imbedded in the base of the skull, is gradually encroached on by the cartilage and bone around it, and ultimately completely absorbed.

2. The Skull.

The skull of the tadpole consists almost entirely of cartilage; none of the bones of the skull, with the exception of the para-sphenoid, appearing until nearly the time of the metamorphosis. In the adult frog, this cartilaginous skull is replaced to a considerable extent by cartilage-bone; while other bones primitively distinct, and probably of dermal origin—the membrane-bones—graft themselves on to it.

The three morphologically distinct elements of which the skull consists (*cf.* p. 43) may with advantage be described separately.

a. **The Cranium** or brain case. This in its fully formed condition is an unsegmented cartilaginous tube, enclosing the brain it is developed as follows:

In the front part of the head a pair of longitudinal cartilaginous bars, the **trabeculæ cranii**, appear in tadpoles of about 10 mm. length: these grow back alongside of the notochord as a pair of horizontal **parachordal** rods.

The hinder ends of the trabeculæ are some little distance apart, and between them is a space in which the pituitary body lies. In front of this pituitary fossa, the trabeculæ unite to form a plate of cartilage, which underlies the anterior end of the brain, and is produced into blunt processes at its outer angles.

The parachordals grow rapidly: they extend inwards so as to

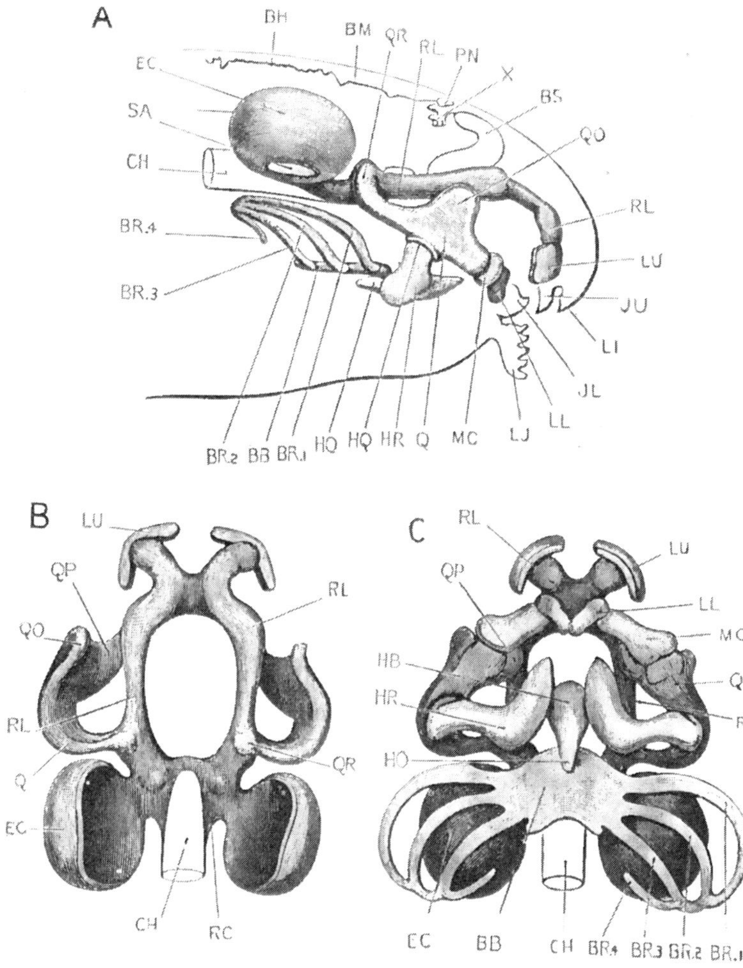

FIG. 34 BIS.

A.—The skull of a 12 mm. tadpole, seen from the right side. ×30.

B.—The same skull from the dorsal surface. The lower jaw and the hyoidean and branchial bars are omitted. ×30.

C.—The same skull from the ventral surface. ×30.

BB, basibranchial; BH, roof of hind brain; BM, roof of mid-brain; BR.1, BR.2, BR.3, BR.4, first, second, third, and fourth branchial bars; BS, cerebral hemisphere; CH, notochord; EC, auditory capsule; HB, basihyal; HO, urohyal; HQ, articulation of ceratohyal with quadrate; HR, ceratohyal; JL, lower jaw; JU, upper jaw; LI, upper lip; LJ, lower lip; LL, lower labial cartilage; LU, upper labial cartilage; MC, Meckel's cartilage; PN, pineal body; Q, quadrate; QO, orbital process of quadrate; QP, palato-pterygoid process; QR, connection of quadrate with trabecula; RC, parachordal cartilage; RL, trabecula cranii; SA, membranous patch in the outer wall of the auditory capsule, in which the stapes is developed at a slightly later stage. X, choroid plexus of third ventricle.

To face p. 141.

meet each other both above and below the notochord, which they now completely surround. The two parachordals soon fuse together to form the **basilar plate**, which, with the trabeculæ, forms a firm cartilaginous floor to the brain case. At their hinder ends the parachordals grow upwards to form the side walls of the cranium, and a little later bend inwards so as to meet each other above the brain, and complete the occipital part of the cranium. Further forwards the pituitary foramen becomes closed by a thin plate of cartilage, and the lateral margins of the parachordals and trabeculæ grow upwards so as to form the side walls of the skull, the roof remaining imperfect in this region.

The first bone to be developed is the parasphenoid. The exoccipitals, the frontals and parietals, which are at first separate, and other bones soon follow; and by the time the metamorphosis is complete and the tail absorbed, all the bones of the adult cranium are present, except the sphenethmoid, which does not appear till some months later.

b. The Sense Capsules. The cartilaginous **auditory capsules** appear in tadpoles of about 12 mm. length as thin shells of cartilage investing the auditory vesicles. They are at first quite independent of the cranium, but before the completion of the opercular folds they fuse with the upgrowing parachordals to form part of the side walls of the skull. The pro-otic appears about the time of completion of the metamorphosis.

The **optic capsules** are thin shells of cartilage, forming part of the sclerotic coats of the eyes. They arise about the same time as the auditory capsules; and, unlike the other sense capsules, they remain distinct from the cranium throughout life, in order to secure mobility of the eyeballs.

The **olfactory capsules** are from their first appearance very closely connected with the anterior ends of the trabeculæ, which grow up between them to form the median vertical internasal septum. They develop later than the auditory and optic capsules.

c. The Visceral Skeleton. This consists of a series of cartilaginous hoops developed within the visceral arches, and forming a framework which surrounds and stiffens the walls of the pharynx. Each hoop consists of right and left halves, which are independent at their dorsal ends, but fused or closely connected ventrally. There are in all six of these hoops or bars forming the oral (mandibular) arch, hyoidean arch, and the four branchial arches respectively; and they develop in order from before backwards.

i. The **oral (mandibular)** bar, which is the largest of the series, lies at first parallel to the others, *i.e.*, perpendicular to the long axis of the body. It very early, however, undergoes important changes, and by the time that the external gills are developed, and before the appearance of the opercular folds, it has altered its direction, and now runs almost horizontally forwards, parallel to and below the trabeculæ.

It soon unites with the trabeculæ, both behind and in front of the eyeball, the latter union being effected by a short transverse bar of cartilage—the **palato-pterygoid.** In front of the palato-pterygoid, the most anterior part of the oral bar becomes segmented off as a short rod of cartilage, which is directed upwards and forwards in the lower lip; it is known as **Meckel's cartilage,** and forms the basis of the lower jaw or mandible. That part of the oral bar with which this segment articulates will give rise to the **quadrate** of the adult. In connection with the lips two pairs of small **labial cartilages** appear, serving to support the horny jaws of the tadpole.

In the later stages the subocular or **quadrate** portion of the oral bar acquires a very close connection at its hinder end with the auditory capsule, and changes its direction, so that in place of running horizontally forwards, it now runs downwards and forwards. This change, which may be described as a rotation backwards of the bar, causes lengthening of the palato-pterygoid bar and of Meckel's cartilage; these latter become respectively the basis of the upper and lower jaws of the tadpole, which are completed later on by the development of the membranous pterygoid, squamosal, maxilla and other bones.

This rotation backwards of the distal end of the quadrate, with corresponding lengthening of the upper and lower jaws, proceeds rapidly during and after the metamorphosis, so that the quadrate, instead of being directed downwards and forwards, soon runs vertically downwards, and later on downwards and backwards as in the adult. (*Cf.* Fig. 10, p. 48.)

ii. The **hyoid bar** also undergoes important changes. At first it is a wide band of cartilage placed nearly vertically in the side wall of the pharynx, immediately behind the oral bar. When the mandibular arch becomes horizontal the hyoid forms a broad stout bar of cartilage, articulating at its upper end with

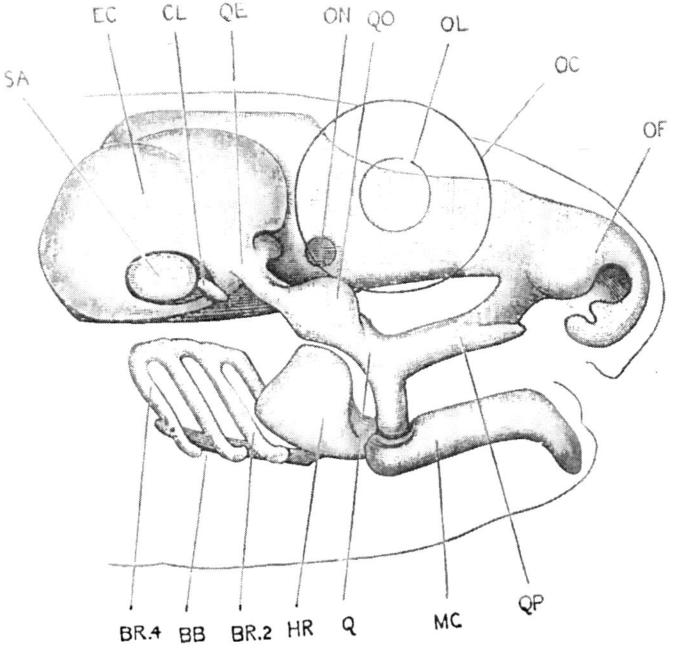

FIG. 34A.—Skull of a tailed frog towards the end of metamorphosis. ×13.

BB, basibranchial; BR.2, BR.4, second and fourth branchial bars; CL, columella; EC, auditory capsule; HR, ceratohyal; MC, Meckel's cartilage; OC, outline of the eye; OC, olfactory capsule; OL, outline of lens; ON, foramen for optic nerve; Q, quadrate cartilage; QE, connection of quadrate with auditory capsule; QO, orbital process of quadrate; QP, palato-pterygoid process; SA, stapes.

To face p. 142.

the subocular part of the oral bar, and connected at its ventral end with the hyoid bar of the other side by a small median **basi-hyal** plate in the floor of the mouth.

At the commencement of the metamorphosis the hyoid bar becomes narrower, and begins to extend upwards towards the auditory capsule; and by the end of the metamorphosis this upper part of the hyoid has become the long slender anterior cornu of the hyoid, which acquires a loose connection at its upper end with the cranium and with the quadrate cartilage.

The development of the **columella** is imperfectly known. It consists of two elements, one of which—the **stapes**—is a small plate of cartilage partially filling a hole, the fenestra ovalis, which appears in the lower and outer wall of the auditory capsule about the time that the opercular folds are growing back over the gills. The other portion of the columella is a small rod, partly cartilage, partly bone, which does not appear till some months after the completion of the metamorphosis, and which fuses with the stapes at its inner end, while its outer end becomes connected with the tympanic membrane (*cf.* Fig. 10, p. 48); this outer element of the columella is commonly regarded as formed from the uppermost part of the hyoid arch, but appears to be really quite independent of it in the frog.

iii. The **branchial bars** are at first simple flattened rods of cartilage, independent of one another, but becoming early connected with a median **basi-branchial** cartilage, which appears in the floor of the mouth between the ventral ends of the first two pairs of bars.

As the hind-legs appear, the branchial bars of each side coalesce with one another both at their dorsal and their ventral ends: they also become strongly curved, and together form a complex basket-work supporting the gills. Later on, as the gills begin to shrink, the branchial bars become more slender: their dorsal ends disappear, while their ventral ends fuse with the basi-hyal and basi-branchial cartilages, and together give rise to the body of the hyoid and its posterior cornua.

N. The Development of the Urinary System.

1. General Account.

The excretory organs of the tadpole, during the early stages of its existence, are the **head kidneys** or **pronephra**. These are a pair of globular organs imbedded in the dorsal wall of the body at its anterior end, immediately behind the constricted neck region (Figs. 33 and 35 KP). Each head kidney is a

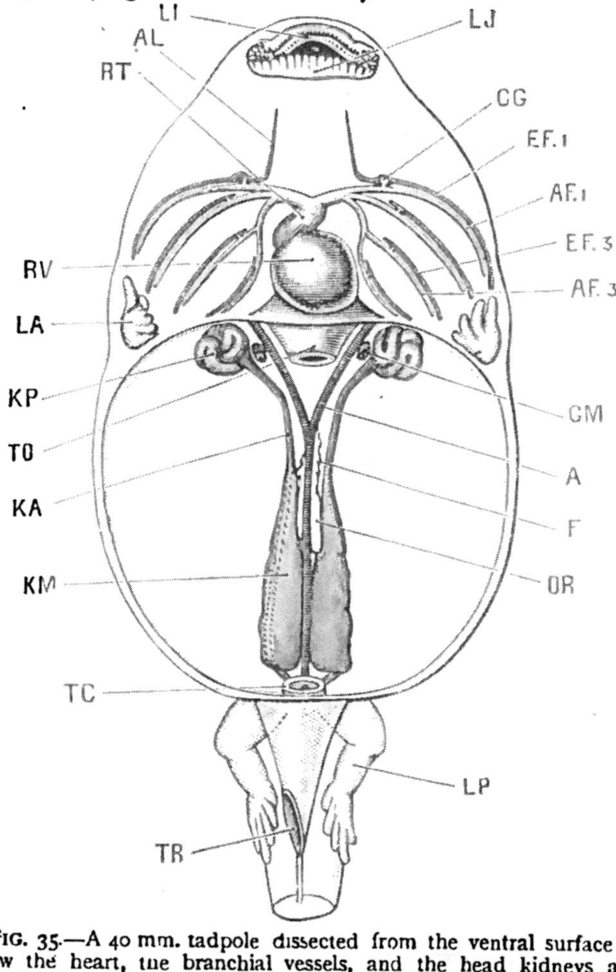

FIG. 35.—A 40 mm. tadpole dissected from the ventral surface to show the heart, the branchial vessels, and the head kidneys and Wolffian bodies. The tail has been cut off. × 5.

A, dorsal aorta; AF_1, AF_3, afferent branchial vessels of first and third branchial arches; AL, lingual artery; CG, carotid gland; EF_1, EF_3, efferent branchial vessels of first and third branchial arches; F, fat body; GM, glomerulus; KA, archinephric or segmental duct; KM, Wolffian body; KP, pronephros or head kidney, now degenerating; LA, fore-limb, still within opercular cavity; LI, upper lip; LJ, lower lip; LP, hind-limb; OR, genital ridge; RT, truncus arteriosus; RV, ventricle; TC, cloaca; TO, œsophagus, cut short; TR, cloacal spout.

convoluted tube with glandular walls, opening into the body cavity by three ciliated mouths or **nephrostomes** (Fig. 33, KS), and continued back along the dorsal wall as the **archinephric** or **segmental duct,** KA, to the hinder end of the body, where it joins with the corresponding duct of the opposite side, and opens into the cloaca.

The head kidneys and their ducts are well developed in the tadpole at the time of hatching: they subsequently increase considerably in size, and are the sole excretory organs of the tadpole during its early stages. In tadpoles of about 12 mm. length the adult **kidneys or Wolffian bodies** (Fig. 33, KM), begin to form in the hinder part of the body as a series of paired tubules, which grow towards and open into the segmental duct. These Wolffian tubules rapidly increase in number, as well as in size and complexity, and become bound together by connective tissue to form the compact **Wolffian bodies** or **kidneys** of the fully formed tadpole (Fig. 35, KM). At the same time the head kidneys diminish in size, and undergo degenerative changes, and by the time of the metamorphosis (Fig. 36) have almost completely disappeared. The Wolffian bodies persist as the kidneys of the frog; and by a series of further changes the ureters and generative ducts of the adult become established.

2. The Head Kidney and its duct.

In tadpoles of about $3\frac{1}{2}$ mm. length, *i.e.*, some time before hatching, a pair of longitudinal grooves appear along the inner surface of the somatopleure, extending from the neck to the hinder end of the body, and lying a little distance to the right and left of the notochord. The lips of each groove soon meet and fuse so as to convert the groove into a tube or duct. The closure of the tube takes place from behind forwards, and at the anterior end is effected imperfectly, three holes or **nephrostomes**, one behind another, being left, through which the tube opens into the body cavity. As the embryo grows, the anterior end of the duct becomes convoluted and twisted on itself to form a ball, the three nephrostomes becoming at the same time lengthened out into short tubes. This convoluted mass is the **head kidney** or **pronephros**. The hinder part of the duct is the **archinephric** or **segmental duct**; it remains straight, or nearly so, and shortly before the tadpole hatches acquires an opening into the cloaca.

At the time of hatching, the excretory organs thus consist on each side of (1) a **head kidney**, which is a convoluted tube lined by a glandular epithelium, and opening into the anterior end of the body cavity by three ciliated openings, the nephro-

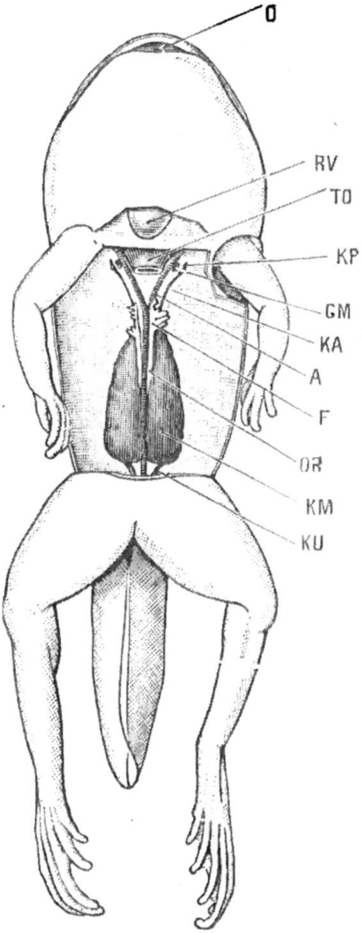

FIG. 36.—A tailed frog, near the close of the metamorphosis, dissected from the ventral surface to show the kidneys and reproductive organs. × 4.

A, dorsal aorta; **F**, fat body; **GM**, glomerulus; **KA**, archinephric or segmental duct; **KM**, Wolffian body; **KP**, head kidney, disappearing; **KU**, ureter; **O**, mouth; **OR**, genital ridge; **RV**, tip of ventricle; **TO**, œsophagus, cut short.

stomes; and (2) the **archinephric** or **segmental duct**, which is the posterior part of the tube, and runs back along the dorsal body-wall nearly straight to the cloaca, into which it opens.

The head kidney is closely surrounded by, indeed almost imbedded in, the posterior cardinal vein (Fig. 37, VC), and it is from the blood of this vein that the epithelial cells of the

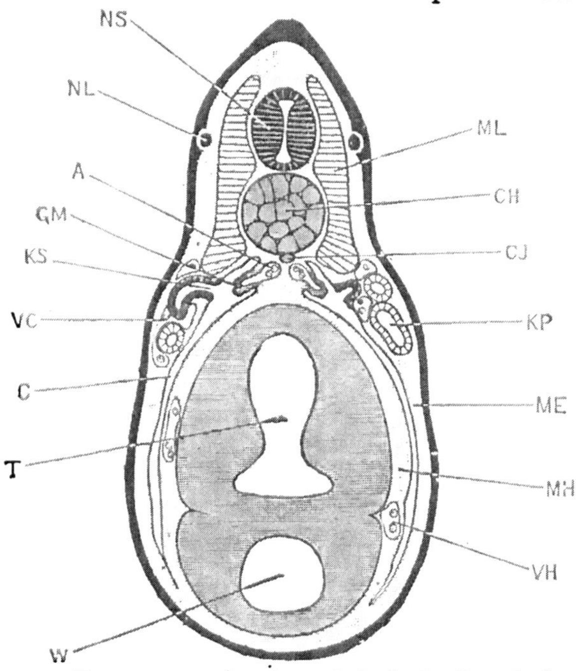

FIG. 37.—Transverse section through the body of a tadpole at the time of hatching; the section passing through the second pair of the nephrostomes, and the third pair of myotomes. × 50. (From Marshall's "Vertebrate Embryology.")

A, aorta; C, cœlom or body-cavity; CH, notochord; CJ, subnotochordal rod; GM, glomerulus; KP, segmental or archinephric duct; KS, second nephrostome of left side; ME, somatopleuric layer of mesoblast; MH, splanchnopleuric layer of mesoblast; ML, myotome; NL, lateral line branch of pneumogastric nerve; NS, spinal cord; T, intestinal region of mesenteron; VC, posterior cardinal vein; VH, hepatic vein; W, liver diverticulum.

head kidney tubules separate the excretory matters, which are then passed down the duct to the exterior.

The head kidney continues to increase in size, the tubules becoming still more convoluted, and lateral diverticula arising from their sides, until the tadpole is about 12 mm. in length, and the hind-limbs are just commencing to appear. It remains stationary for a time and then, in tadpoles of about 20 mm. length, begins to degenerate: the tubules become obstructed; some of them become collapsed, others for a time irregularly dilated: the whole organ steadily diminishes in size, and in tadpoles of 40 mm. (Fig. 35, KP) is not more than half its former size. It now shrinks rapidly, and at the time of the metamorphosis (Fig. 36, KP) has almost disappeared, all three nephrostomes having closed up, and the organ being reduced to a few small pigmented and irregularly twisted tubules, which have separated from the duct, and which soon disappear completely.

Opposite the head kidney an irregular sacculated outgrowth, the **glomerulus,** arises from the aorta on each side (Figs. 31 to 37, GM): this appears first about the time of hatching, and its development keeps pace with that of the head kidney. It lies immediately opposite the nephrostomes, and very close to these, though not touching them. It begins to diminish in size about the same time as the head kidney. At the time of the metamorphosis (Fig. 36, GM) it is very small, and after the first year it can no longer be recognised. Its close relation to the head kidney, and the fact that its growth and subsequent degeneration keep pace with those of the head kidney, point to a close physiological connection between the two organs, though it is not easy to imagine what precise function the glomerulus subserves.

3. The Wolffian Body.

The Wolffian body, or kidney, first appears in tadpoles of from 10 to 12 mm. in length. It arises on each side as a series of small solid masses of mesoblast cells lying along the inner side of the segmental duct, between this and the aorta (Figs. 33 and 35). They develop from behind forwards, the hindmost pair being a short distance in front of the cloaca, and the most anterior ones about three segments behind the head kidney.

These solid masses soon become elongated into twisted rods, which then become tubular, and growing towards the segmental duct meet and open into it. At their opposite ends these **Wolffian tubules,** as they are termed, dilate into bulb-like expansions, which become doubled up by ingrowth of little knots of bloodvessels, derived from the dorsal aorta, and so form Malpighian bodies. From the necks of the Malpighian bodies, short solid rods of cells grow towards the peritoneal epithelium and fuse with it. These rods soon become hollow, and open into the body cavity by ciliated funnel-shaped mouths or nephrostomes: their opposite ends break away from the Wolffian tubules and open directly into the renal veins on the ventral surface of the kidney. The Wolffian tubules rapidly increase in number; they also branch freely, and so give rise to a complicated system of glandular tubules, which, when bound together by bloodvessels and connective tissue, form the Wolffian body or kidney of the frog. The nephrostomes persist: and in the adult frog as many as 200 or more are present on the ventral surface of the kidney, as minute funnel-like ciliated openings, leading by short tubes into the renal veins.

4. The Wolffian and Müllerian ducts.

So far we have only described one duct on each side, the segmental duct, which acts as the excretory duct first of the head kidney, and then of the Wolffian body as well. We have now to consider in what way the ureters and generative ducts of the adult frog are formed.

About the time of the metamorphosis the head kidney, which has become rudimentary, separates completely from the duct, which now ends blindly a short distance in front of the Wolffian body.

A little later, after completion of the metamorphosis and entire disappearance of the tail, this anterior end of the segmental duct, in front of the Wolffian body, becomes divided somewhat obliquely into two; an anterior part, which is now isolated from the Wolffian body, and will be called the **Müllerian duct**; and a posterior part, the **Wolffian duct**, which is simply the posterior part of the original segmental duct, and receives the Wolffian tubules of the kidney.

The **Müllerian duct** becomes connected in front with the peritoneal epithelium, and acquires an opening into the anterior end of the body cavity. At its hinder end it grows back along the outer side of the Wolffian duct to the cloaca, into which it opens. So far the changes are the same in both sexes. In the male frog the Müllerian duct persists in this condition throughout life, and may be recognised as a slender longitudinal streak lying in the thickness of the peritoneum a short distance to the outer side of the kidney, and extending some distance in front of it. In the **female frog** the **Müllerian duct** becomes the oviduct, the anterior opening being carried forward first as a groove, and then by closure of the lips as a tube, to the position characteristic of the peritoneal opening of the adult oviduct; while the posterior part becomes greatly convoluted and acquires thick glandular walls: the hindmost part of the oviduct remains thinner walled, but of much greater capacity.

The **Wolffian duct** becomes in both sexes the **ureter.** In the **female frog** it undergoes no further change of importance. In the **male frog** the hinder end of the Wolffian duct becomes dilated into a much-branched glandular enlargement, the **vesicula seminalis.**

5. The **Vasa Efferentia.**

In both sexes at an early stage, as the Malpighian bodies are forming in the Wolffian body, those nearest to the genital ridges give off tubular branches from their capsules into the ridges.

In the **female frog** these tubules are said to expand very greatly, and to give rise to the chambers or cavities present in the adult ovary: but the point is not established with certainty.

In the **male frog** these tubules become the vasa efferentia: they become connected with the spermatic tubules, and, as at their other ends they open into the Wolffian tubules, they form passages along which the spermatozoa can get from the testis to the Wolffian duct or ureter, and so out.

CHAPTER IX.

ELEMENTARY HISTOLOGY.

When examined under the microscope, all the different tissues and organs of the body are found to consist of elementary bodies called **cells** and of an **intercellular substance**, connecting the several cells together; in much the same way as a wall is built of bricks cemented together with mortar. These cells, of which a white blood corpuscle is a typical example, vary much in shape, size, and structure in different tissues, but are to be considered as fundamentally equivalent to one another. The intercellular substance varies very much in quantity; it may be almost absent, so that the several cells are practically in contact with one another; or it may be so abundant as to separate them widely; it is to be viewed as formed by the cells, and, therefore, as secondary in importance to these.

When drawing histological preparations, it is well to look out for, and draw, a few red blood corpuscles, to the same scale as the rest of the drawing. The blood corpuscles form most useful standards of measurement, as their dimensions are already known (p. 37).

A. Epithelium.

Epithelium consists of cells placed side by side so as to form layers, which form the surface covering, or epidermis, of the body, and line the alimentary canal, the blood vessels, and the various internal cavities of the body. It may be defined as a continuous sheet of cells lining a free surface. At the external apertures of the body, the epidermis is directly continuous with the epithelial lining of the internal cavities.

The layers may be one or more cells in thickness; in the former case the epithelium is said to be **simple**, in the latter **stratified**.

Epithelium is of different kinds, according to the shape and structure of its component cells.

I. Squamous Epithelium. In this the component cells are flattened parallel to the surface they cover; if the epithelium is stratified, the flattening is most marked in the superficial cells.

a. Isolated Cells.

Scrape gently the inside of your cheek with the handle of a scalpel, and put the scrapings on a slide; cover, and examine with a high power; draw, showing the following points:

i. The cells are large, flattened and scale-like in shape, often slightly curled up at their edges.

ii. The **nucleus** is oval and granular, and lies near the middle of the cell; it may be rendered more distinct by acetic acid or magenta.

b. Cells in situ; cast skin of newt.

Take a small piece of the prepared specimen, which has been stained in hœmatoxylin, and then, after treatment with alcohol, cleared with oil of cloves. Mount the specimen in balsam; cover, and examine with the high power.

i. The cells are flattened, and fitted together at their edges, like a mosaic, to form a continuous layer. Each cell has a large nucleus near its centre.

II. Columnar Epithelium. This consists of elongated rod-like cells, placed vertically to the surface on which they rest. If a columnar epithelium is stratified the columnar character is most marked in the superficial cells.

a. Isolated cells: from the small intestine of the frog; isolated by maceration for 24 hours in Ranvier's alcohol, and stained with picro-carmine.

Mount a drop of prepared specimen in glycerine; paint a ring of cement round the cover-glass; and examine with the high power.

i. The cells, which often remain side by side in little groups, are columnar in shape, with nuclei near their inner or deeper ends.

b. Cells in situ.

Take a prepared section of dog's stomach which has been stained, and then cleared in oil of cloves. Mount in balsam, and examine with the high power.

i. The superficial layer consists of long narrow columnar cells, packed together side by side, with nuclei at their inner or deeper ends.

III. Ciliated Epithelium. In this the cells, which are usually columnar, bear at their free ends tufts of exceedingly fine hair-like processes—cilia—which, when living, exhibit active lashing movements.

a. Isolated cells. From trachea of rabbit: isolated by maceration for 24 hours in Ranvier's alcohol; stained with picro-carmine, and scraped into glycerine.

Mount a small drop of the prepared specimen in glycerine; paint a ring of cement round the cover-glass; examine with the high power, and note:

i. The shape of the cells: their nuclei; and the tuft of cilia at one end of each cell.

b. Cells in situ: ciliary movement.

Snip off a small piece of epithelium from the roof of the mouth of a freshly killed frog, near the eyeball; mount in normal salt solution, and add a small drop of gamboge water to render the movements more clearly visible; examine with the high power: note:

i. The currents due to the ciliary motion.

ii. The movements of the individual cilia: best seen when the specimen is beginning to die, and the movements to slacken in speed.

IV. Stratified Epithelium. This is characterised by the epithelium being several cells in thickness.

Take a prepared section of œsophagus of rabbit, or of conjunctiva of rabbit or pig, which has been hardened in chromic acid, stained, and cleared in oil of cloves. Mount in balsam, examine with the high power, and note:

i. The stratification of the epithelium.

ii. The transition from the deeper spherical or columnar cells to the superficial squamous cells.

B. Glands.

A gland consists essentially of a layer of epithelial cells secreting some special fluid. The epithelial surface may be flat, but is more usually folded or pitted, often in a very complicated manner, so as to increase the extent of the secreting surface.

I. Simple Glands. In simple glands the epithelial surface is increased by simple pit-like depressions, whose mouths serve to discharge the secretion on the free surface.

Take a prepared section of large intestine of rabbit which has been hardened in chromic acid, stained, and cleared in oil of cloves. Mount in balsam, and examine first with the low power, then with the high. Note the following points:

i. The glands are simple tubular depressions of the surface.

ii. The glandular epithelium lining the pits is a single layer of short columnar granular cells, many of which are swollen to form **goblet cells.**

II. Compound Glands. In compound glands each depression instead of being a simple pit is itself subdivided or branched, often in a very complicated manner. There are two chief varieties: (1) **tubular glands,** in which the several subdivisions are tubular, and of tolerably uniform diameter throughout: and (2) **racemose glands,** in which the blind ends of the pits are dilated into globular chambers or alveoli, to which the special glandular epithelium is usually confined.

a. **Compound tubular glands.** *Take a prepared section of kidney of frog. Examine under a hand lens, and draw on a large scale. Notice that the dorsal surface may be distinguished from the ventral.*

i. The **tubular gland cavities** are cut at various angles. If cut transversely a tube appears as a circular ring: if cut obliquely, as a more or less elongated elliptical ring: if cut longitudinally, as two parallel rows of epithelial cells.

ii. The **gland cells** form a single layer of cubical granular cells, lining the tubes.

iii. The **Malpighian bodies** are spherical dilatations on the tubes, into which project little knots of capillary

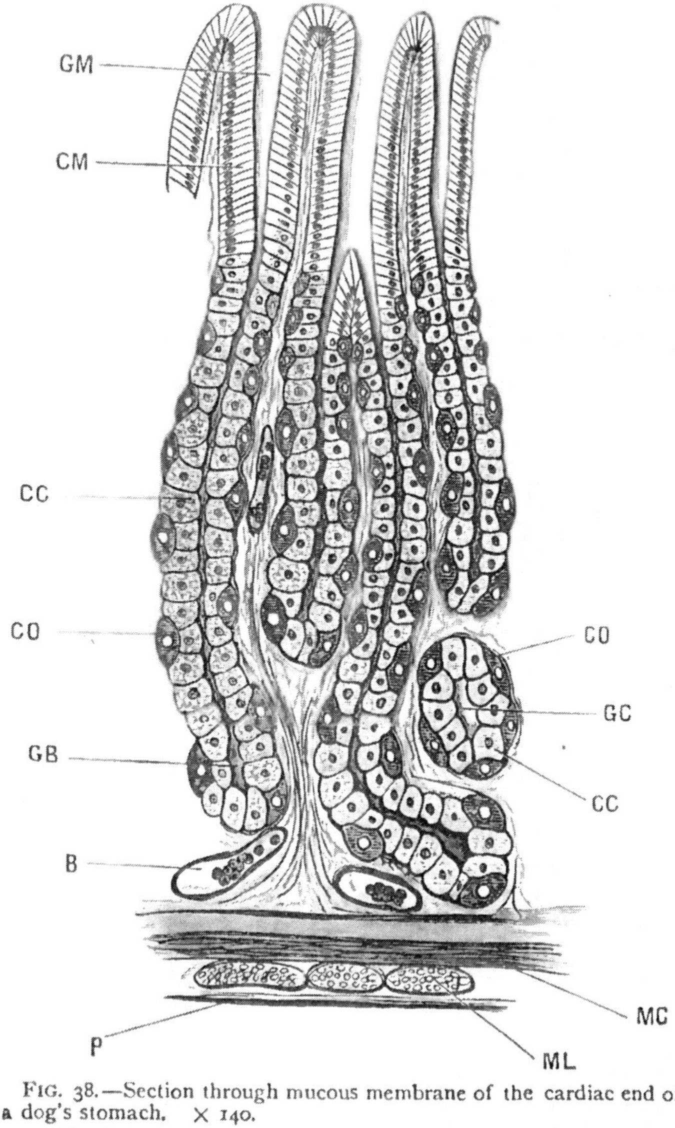

FIG. 38.—Section through mucous membrane of the cardiac end of a dog's stomach. × 140.

B, bloodvessel; CC, cubic or peptic cells; CM, columnar cells; CO, ovoid cells; GB, fundus or bottom of gland cavity; GC, gland cavity cut across; GM, mouth of gland; MC, circular muscle fibres; ML, longitudinal muscle fibres; P, connective tissue layer between the mucous membrane and the outer muscular walls of the stomach.

bloodvessels. Their structure is most readily made out in specimens in which the bloodvessels have been injected with a coloured substance to make them more distinct.

III. Gastric Glands. The glands of the stomach are well adapted for a more minute examination of the histology of glands.

Examine again with a high power the section of the cardiac end of the dog's stomach already used for columnar epithelium.

1. **Characters of the glands.** The gastric glands are good examples of simple or slightly branched tubular glands. They are deep, but very narrow, cylindrical pits, imbedded vertically in the wall of the stomach, with their open mouths discharging into its cavity. The glands are lined by epithelial cells, and are set very close together side by side. In the microscopical sections, some of the glands may be seen cut along their entire length; but in most cases, owing to the glands being not quite straight, or the plane of section being oblique to the surface of the stomach, the tubes will be cut more or less obliquely, or even transversely.

2. **Characters of the gland cells.** There are three distinct kinds of epithelial cells found at different parts of the length of the gland.

   i. **Columnar** cells, arranged in a somewhat radiate manner round the mouths of the glands, and extending a short way down the tubes.

   ii. **Cubic cells,** or peptic cells, lining the deeper parts of the glands and the greater part of their length: these are cubical granular cells with centrally placed nuclei.

   iii. **Ovoid cells**: large oval cells with large nuclei: these are less numerous than the other two forms, and occur most abundantly a short way below the mouths of the glands. They lie along the sides of each gland, outside the cubical cells, and are said to secrete the acid of the gastric juice.

IV. Glands of Frog's Skin. Sections of the frog's skin are admirably suitable for histological purposes, as they show a collection of typical tissues.

Prepare sections thus: Take a fresh-caught summer frog; chloroform it; cut off a small piece from the skin of the back; fix with corrosive acetic acid; stain with hæmatoxylin, or stain the sections on the slide.

The **Epidermis** is stratified, pitted into two sorts of glands.

1. **Granule glands.** Very large, probably poisonous; sacs without an opening to the duct. They disappear in captivity.
2. **Slime glands.** Flask-shaped sacs that produce the mucus with which the body is covered and safeguarded from the germination of parasites.

The pigment-cells, muscle-cells, and blood-cells are excellently shown.

C. Muscle.

In muscular tissue the component cells are much elongated and, in the higher forms, very highly specialised. Muscular tissue is of two kinds: (1) **striated, or voluntary,** of which all muscles that are under the control of the will consist: and (2) **non-striated or involuntary,** forming those muscles over whose contractions the will has no direct control. The muscular tissue of the heart, which though involuntary is striated, forms the chief exception to this rule.

I. Striated, or Voluntary Muscle.

a. **Crab's muscle.** *Shred in glycerine a small piece of crab's muscle that has been hardened in alcohol; cover, and examine with both low and high powers: note:*

i. The elongated fibres of which the muscle consists. Each fibre is a single cell, and is enclosed in a delicate elastic sheath—the **sarcolemma**—which will be visible in but few cases; it is most readily seen at places where the fibre has been torn across.

ii. The alternate light and dark bands with which the muscle fibres are marked transversely, and from which the name, striated muscle, is derived.

iii. The readiness with which the fibres split up longitudinally into **fibrils**.

b. **Frog's muscle.** *Shred gently a piece of fresh frog's muscle in normal salt solution: cover, and examine with the high power: note:*

i. The **transverse striations**.

ii. The **sarcolemma**: best seen by slightly crushing the specimen.

iii. The **nuclei** in the fibres: seen on addition of acetic acid.

II. Non-striated, or Involuntary Muscle.

Take a prepared specimen of frog's bladder which has been macerated in Ranvier's alcohol for 24 hours; pencilled with a fine brush to remove the epithelium of the inner surface; stained, and cleared with oil of cloves. Mount in balsam, and examine with low and high powers: note:

i. The bands of muscular fibre.

ii. The formation of each band by a number of elongated, fusiform, nucleated muscle-cells.

iii. The absence of transverse striation in the muscle.

D. Connective Tissues.

Under the name "connective tissue" are included various tissues whose functions are mainly passive, and which serve to support, strengthen and bind together the various organs and parts of the body. Histologically the connective tissues consist of elements of four kinds, united together in very varying proportions in different situations: (1) white fibrous tissue, (2) yellow elastic tissue; (3) connective tissue corpuscles, which are comparatively slightly altered cells, usually branched; and (4) ground substance, or intercellular substance.

I. **White Fibrous Tissue.** This consists of a number of fine transparent fibres of a more or less cylindrical shape, and with a very characteristic wavy outline; between the fibres are connective tissue cells, usually in small numbers. The fibres are arranged side by side in bundles, and each fibre presents a number of longitudinal fibrillar striations. The cellular origin of white fibrous tissue is difficult to determine. The fibres are believed to be formed by modification of the intercellular matrix rather than from the bodies of the cells themselves.

a. **Tendon of rat's tail.** *Pull out a small piece of tendon from the tail of a rat; place it on a slide in a drop of normal salt solution; spread it out with needles, cover and examine with low and high powers: note:*

i. The **fibres**, with wavy outlines.

ii. The **fibrillæ**, indicated by longitudinal wavy striations within the fibres.

Add a drop of acetic acid to the preparation: note that

iii. The fibres swell up and become transparent.

iv. Longitudinal rows of **tendon cells**, with nuclei, become visible between the fibres.

II. **Yellow Elastic Tissue.** This consists of fine branching homogeneous fibres, with great power of resisting chemical reagents; the fibres are formed from an intercellular matrix, and not from cells directly.

a. **Ligamentum nuchæ of ox.** *Tease finely a small shred in water; examine with low and high powers: note:*

i. The branching fibres, with very sharp outlines.

ii. The tendency of the branches to anastomose with one another and so form networks.

iii. The tendency of the fibres and branches to curl up at their broken ends.

Add a drop of acetic acid: note that

iv. No alteration whatever is produced in the fibres.

v. No nuclei appear.

III. **Areolar tissue.** This is a meshwork composed of both white fibrous and elastic tissues.

a. **Subcutaneous tissue of mammal.** *Take a freshly killed rat, and snip off a small piece of the loose fibrous tissue which connects the skin with the subjacent parts; stretch it till quite flat with a pair of needles, breathing constantly upon it to make it adhere to the slide; cover, and examine with low and high powers: note:*

i. The meshwork, composed of white fibrous tissue with wavy outlines, mingled with which are branched elastic fibres.

Add acetic acid: note that

ii. The white fibrous tissue swells up and becomes transparent.

iii. The elastic tissue is unaltered.

iv. Connective tissue corpuscles, with nuclei, become visible.

IV. **Adipose tissue.** This consists of a fine network of vascular connective tissue, in the meshes of which are fat cells, *i.e.*, connective tissue corpuscles in which large quantities of fatty or oily matter have accumulated.

a. **Omentum of rabbit or kitten.** *Mount a small piece of fresh omentum in normal salt solution; protect it from* *the pressure of the cover glass; examine with low and high powers: note:*

i. The vascular connective tissue meshwork, in which lie the fat cells.

ii. The **fat cells**: large, spherical, or from mutual pressure polyhedral, cells; distended with fatty matter, and with their nuclei near the surface.

b Osmic Acid specimen.

Note the reduction of the osmic acid by the fat, which becomes stained a dark brown or black colour.

E. Cartilage.

In cartilage or gristle the intercellular substance, which in most other tissues is only present in small quantity, is greatly increased so as to far exceed in bulk the cells which it connects together. The intercellular substance forms a dense translucent matrix resembling an extremely stiff jelly, in which are imbedded the cartilage cells, either singly or in groups. In young cartilage the intercellular substance is much less abundant, and the cells consequently closer together than in older or more mature specimens.

Cartilage when free from other tissue is called hyaline cartilage, from the clear or glassy appearance of the matrix, in contradistinction to fibro-cartilage, in which the matrix is fibrous from admixture with white fibrous or elastic tissues.

I. Hyaline Cartilage.

a. Cartilage of newt. *Take a small piece of cartilage from the shoulder girdle of a newt; scrape away gently any muscle or other tissue that may adhere to it; mount in normal salt solution, and examine with low and high powers.*

i. The **intercellular matrix** is either hyaline or faintly granular.

ii. The **cartilage cells** are imbedded in the matrix; each cell is nucleated, and occupies a cavity or lacuna in the matrix. In places the cells are in groups of twos or fours owing to recent division.

Wash the specimen thoroughly in water; stain with carmine, and mount as a permanent preparation in glycerine; examine with the high power, and note that

iii. The cell nuclei are stained deeply, and the matrix very slightly; the layer of matrix immediately surrounding each cell—the **capsule**—stains more deeply than the other parts.

b. **Articular cartilage.** This forms caps covering the ends of those bones which fit together to form movable joints; the caps act as elastic cushions to break the force of shocks.

Mount in balsam a prepared section of articular cartilage from the head of the femur, the section being made perpendicular to the articular surface; examine with low and high powers.

i. The **matrix** is hyaline or faintly granular.

ii. The **cartilage cells.** Towards the free surface the cells and cell groups become gradually flattened, and arranged parallel to the surface.

F. Bone.

Bone consists of a dense fibrillar intercellular matrix, in which are imbedded cells which lie in cavities connected with one another by fine branching canals. The matrix is richly impregnated with inorganic salts, chiefly phosphate and carbonate of lime, which form about two-thirds by weight of the substance of the bone, and give it its great hardness and strength. The matrix, with its contained bone-cells, is arranged in concentric layers or lamellæ, around tubular passages, the **Haversian canals,** in which lie the bloodvessels, which penetrate the bone in great numbers. A Haversian canal with its contained bloodvessels, and its surrounding layers of matrix and cells, are together spoken of as a **Haversian system.**

1. *Examine with both low and high powers prepared transverse sections of a long bone.*

i. The **Haversian systems** form the greater part of the bone, and are readily recognised by the concentric arrangement of the lamellæ, and the central canals.

ii. The **interstitial lamellæ** fill up the spaces between the Haversian systems. They form parts of circles which are in many cases of much larger radius than the circles of the Haversian systems.

iii. The **lacunæ** are the spaces in the matrix in which the bone-cells lie. In sections of dried bone the lacunæ appear black, through being filled either with air or with dirt.

iv. The **canaliculi** are very fine branching canals connecting the lacunæ together; they are occupied while the bone is living by branching processes of the bone-cells. At the outer part of each Haversian system, some of the canaliculi are looped, opening at both ends of the loop into the same lacuna.

v. The large central **medullary cavity** of the bone is occupied during life by the **marrow**, which consists of adipose tissue, with very numerous bloodvessels and large nucleated reddish marrow cells.

vi. The **peripheral or circumferential lamellæ** are a series of concentric lamellæ parallel to the surface of the bone, and forming its most superficial layer.

vii. The **perimedullary lamellæ** are a series of concentric lamellæ lining the central medullary cavity of the bone.

CHAPTER X.

CELL DIVISION: DEVELOPMENT OF GERM-CELLS.

THE great majority of cells possess the faculty of division. The division by simple fission of the nucleus and cytoplasm (Direct Division) is a rare mode. The most general method is one of much greater complexity, and is accompanied by the formation of threads in the nucleus, and by movements of those threads (Indirect Division). Hence the division of most cells is known as **mitosis** or **karyokinesis**, in allusion to the nuclear thread and nuclear movements that accompany the process.

Before the process of indirect division can be understood it is necessary to explain certain details of cell-structure. Every cell consists of a corpuscle of protoplasm differentiated into two substances—one nuclear, the other cytoplasmic. The nuclear substance is generally in the form of a definite body, the nucleus, which is imbedded in the surrounding cytoplasm. The nucleus is not uniform in structure, but exhibits a spongy character and consists of a substance called linin, and holds a viscous nucleoplasm in its meshes. These substances, the linin and the nuclear sap or nucleoplasm, do not stain with logwood dyes, hence they are called the achromatic elements of the nucleus. The linin network is, however, concealed by a closely applied covering of granules, thickened here and there at the nodes or knots, and this covering stains strongly with most dyes, and hence is called chromatin. When the knots of this chromatic network are very marked they are called nucleoli, but there are also one or more spherical bodies lying in the nuclear sap, and these also are called nucleoli.

When a cell is about to divide, and throughout life in

some cases, an extremely minute body is found outside but close to the nucleus. This body consists of a central granule, and is usually surrounded by rays. The granule is called the "**centrosome,**" and the rays the "**aster.**"

I. The Changes during Cell Division.

Take a prepared and stained section of the newt's testis, or of some other actively dividing tissue, in order to see the structures mentioned below. Owing to the fact that the plane of this section cuts many of the cells at very different angles, the nuclear figures even at the same phase of division will appear very different. Examine and draw under a high power.

1. Prophase.

A. The nuclear changes.

The chromatin substance increases in staining power, loses its net-like arrangement, and gives rise to a definite number of separate intensely staining bodies known as **chromosomes** (Fig. 39). In many cells the chromatin takes the form of a spirally wound thread (**spireme**) before it is broken up into chromosomes.

A striking fact about the chromosomes is that their number is constant in every somatic tissue-cell of a given species of animal or plant, and that this number in all organisms that reproduce sexually is an even number.

B. The achromatic figure.

While the chromosomes are arising, a series of changes is taking place at the periphery of the nucleus. The centrosome of the undivided cell splits into two, and the two centrosomes move apart as a spindle of delicate fibres arise between them. Meanwhile the outer wall of the nucleus has become less distinct, and the nucleoplasm is mingling with the cytoplasm.

Around each centrosome a number of excessively delicate radiating processes are forming—the astral rays—so that the spindle appears to be giving off radiations at its ends. This figure is the **amphiaster** or achromatic figure, and occupies the position held formerly by the nucleus.

The chromosomes now arrange themselves about a plane (equatorial plane) at right angles to the axis of the spindle, thus forming the equatorial plate.

The entire figure thus formed by the achromatic spindle and the chromosomes is known as a **mitotic figure.**

2. Metaphase.

The preceding changes are preparatory to division, which now begins by a longitudinal splitting of each chromosome into two equal halves. In some cases the chromosomes split only after they have taken up their position at the equator of the spindle. This stage of division is known as the *metaphase.* In other cases the division of the chromosomes occurs before the equatorial grouping of these bodies. In either case the original chromatin becomes divided into exactly equivalent halves.

3. Anaphase.

After the chromosomes have split, one half of each chromosome travels along the spindle-fibre to one pole of the spindle, the other half travels to the opposite pole. The result is a group of daughter chromosomes at each pole of the spindle.

4. Telophase.

The entire cell now divides into two by a plane passing through the equator of the spindle, each of the daughter cells receiving an equal number of similar chromosomes, half of the spindle and one of the asters with its centrosome.

The chromosomes now lose their individuality and reconstruct a fresh nucleus. The spindle disappears together, in most cases, with the centrosomes and asters. In a few cases, however, the latter persist.

This process often takes place rapidly (in about one or two hours), and usually during the night or early morning hours. Between one division and the next there is a long pause. Hence in any section the majority of cells do not show the process of division unless the tissue has been prepared during the critical period.

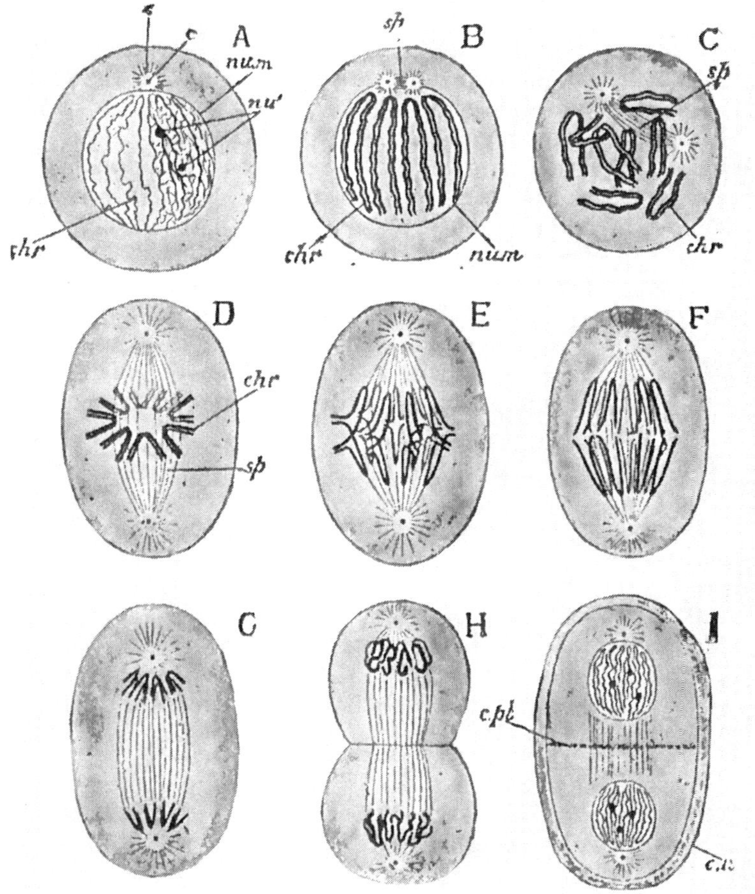

FIG. 39.—Diagram of the phases of cell-division by mitosis. (After Parker.) Highly magnified.

A, cell about to divide; B, division of the chromatin into chromosomes. The centrosome has divided into two; C, formation of a spindle (*sp.*) between the two centrosomes; D, equatorial arrangement of the chromosomes; E, commencing migration of the split chromosomes towards the poles of the spindle; F, G, H, formation of daughter-nuclei and division of the cell; I, formation of a cell-plate between the daughter-cells.

c, centrosome; *chr*, chromosomes; *c.pl*, cell-plate; *cw*, cell-wall; *nu'*, nucleoli; *num*, nuclear membrane; *s*, aster; *sp*, spindle.

II. Development of the Germ-Cells.

In the body of all multi-cellular animals there are produced cells which have for their function the continuance of the species. These cells, known as **germ-cells** or **gametes**, are of two kinds: spermatozoa or male gametes, and ova or female gametes. As a general rule ova cannot give rise to new individuals until they have been fertilised by spermatozoa; and this process of fertilisation again is only possible after both gametes have attained maturity.

A mature ovum consists of a usually spherical mass of cytoplasm enclosing a nucleus. The cytoplasm generally possesses a sponge-like consistency, and contains in its meshes food material and frequently pigment. The whole ovum is enclosed in an elastic membrane, the *vitelline membrane.*

Examine the ova of the earthworm or sea-urchin. Also draw sections of the ovary of queen wasp, frog, or rabbit.

A mature spermatozoon is always exceedingly small and usually slender. It consists, in most animals, of a minute knob or rod carrying a delicate whip-like filament. During life this "tail" is capable of rapid vibratile movement, the tendency of which is to cause the spermatozoon to swim with the knob forwards. Hence this thickened end is called the "head." Between the head and the whip-like tail is a minute thickening —the middle piece—which is the root of the tail.

Examine mature spermatozoa (from the vesicula seminalis of a frog) in salt solution. Draw under the high power and notice the movements due to flexion of the tail.

Maturation of the Germ-Cells. Reduction of the Chromosomes.

A. Oogenesis.

The primary ova are found at a very early stage of animal development, in many cases before birth, and in a few cases from the first cleavage of the egg into two cells, one of which gives rise to the body and the other to the germ-cells.

The primary ova are found in the peritoneum along each

side of the aorta. Here they divide many times to form **oogonia**. Their mode of division is the indirect one (karyokinesis), and is accompanied by the formation of a mitotic figure as described on pp. 165-6 for a tissue-cell. Each oogonium contains the same number of chromosomes as does each tissue-cell. The whole mass of such oogonia, together with the blood-vessels, connective tissue, nerves, and the peritoneal covering associated with them, constitute the ovary.

These germ-cells, however, are not mature, nor are they capable of giving rise to a new generation of individuals until they have passed through (*a*) a period of growth, (*b*) a process of maturation, and (*c*) a process of fertilisation.

Shortly before the first batch of eggs is laid, a certain number of oogonia begin to store up yolk in their cytoplasm. This stage is the period of growth. In the frog about 3000 oogonia undergo this change during the winter. Such large immature germ-cells may be called **primary oocytes,** whilst the remaining oogonia are reserved for a subsequent spawning period.

The next stage is occupied by the process of maturation or ripening. This process consists in the extrusion of nuclear and cytoplasmic substances from the oocyte in the form of two corpuscles known as the **polar bodies.** The nucleus of the oocyte moves to the surface. A mitotic figure is developed which differs, however, in an important aspect from the mitotic figure of the tissue-cells and of dividing oogonia. The chromosomes are fused in pairs so that only half the full number is present. In other respects the division of the spindle and chromosomes takes place as described on p. 166. Each double chromosome divides transversely, one half going to each pole of the spindle. As the spindle with its attached chromosomes divides, one half passes to the exterior of the oocyte together with a small amount of cytoplasm and is pinched off as a small cell—the **first polar body.** The other half of the spindle now retires into the interior of the oocyte, and its chromosomes are reconstituted into a fresh nucleus.

After an interval this nucleus repeats the process, and the **second polar body** is extruded. This has again the reduced number of chromosomes. In the case of certain animals the first polar body meantime has divided into two. The remaining nucleus again retires into the interior of the egg-cytoplasm to await the advent of the spermatozoon. The egg is now mature, and capable of fertilisation. It contains half the number of chromosomes that were in the oogonium.

Examine sections of the ovary of Ascaris with the high power in order to see the formation of the polar bodies. Sections of the ovary of the frog, rabbit, or mouse also show the growth of the oocyte.

B. Spermogenesis.

The primary sperm-cells can be recognised at an early stage of development, in this respect agreeing with the primary ova. These **spermogonia,** as they are called, become imbedded in the peritoneum on either side of the aorta. Over this area the peritoneum develops a series of tubular folds which constitute the testis. In this position the spermogonia divide either by the direct method, or more generally as described for a tissue-cell on p. 165. The full number of chromosomes is present. After a considerable number of such divisions the resulting cells cease to multiply and commence to grow. They may now be called **primary spermocytes.** Each spermocyte is capable of giving rise to four spermatozoa. The first step in the process is the division of a spermocyte into two cells. This division always takes place by karyokinesis, and is accompanied by a reduction in the number of chromosomes to half the initial number (Fig. 40). A nuclear figure is formed, but as the chromatin breaks up into chromosomes these bodies fuse in pairs to form double chromosomes. Each spermocyte behaves like an oocyte, two equal **secondary spermocytes** are formed, and these rapidly dividing again give rise to four **spermatids,** each of which has half the number of chromosomes present in the primary spermocyte. These four spermatids correspond with the ripe ovum and its three polar bodies.

The spermatids are, however, unripe spermatozoa. To become mature they have to go through a complicated development. Each of these cells contains a centrosome and a nucleus imbedded in a spherical mass of cytoplasm. The first change is the outgrowth of a delicate thread from the centrosome.

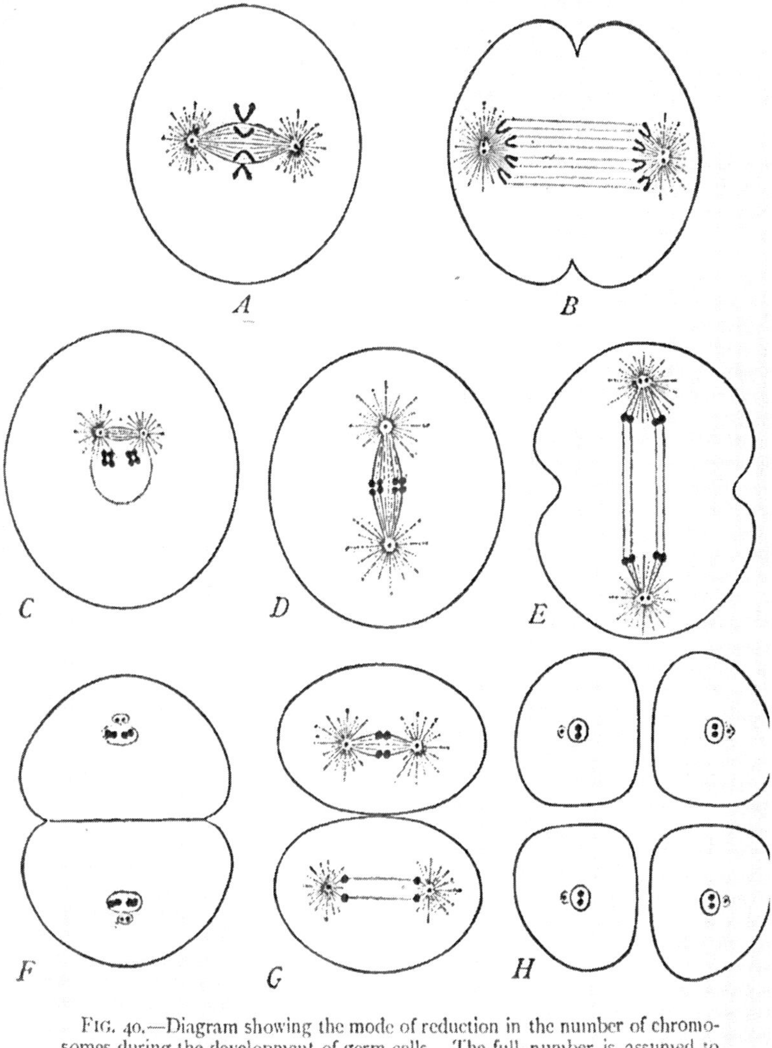

FIG. 40.—Diagram showing the mode of reduction in the number of chromosomes during the development of germ-cells. The full number is assumed to be four. (After Wilson.)

A, B, division of an oogonium or of a spermogonium, showing the full number of chromosomes. C, Origin of the double chromosomes in the primary spermocyte. D, E, F, division of this cell into two secondary spermocytes, each containing two double chromosomes. G, division of the secondary spermocytes to form four spermatids, H, each with two single chromosomes and a centrosome.

The mode of reduction is essentially the same in the maturation of an egg, but of the four cells that result, the three polar bodies are much smaller than the egg.

As this elongates it carries with it an extremely delicate layer of cytoplasm. The thread becomes contractile, and in this way the tail of the spermatozoon is formed, the centrosome remaining at the root of the tail as the middle piece. The head of the spermatozoon arises chiefly from the condensed nucleus of the spermatid, and possesses a thin covering of cytoplasm.

Examine prepared sections of the testis of the newt, frog, or rat.

C. Formation of the Zygote.

If a mature ovum, as described above, be fertilised by a ripe spermatozoon, the following changes take place. By the aid of its vibratile tail the spermatozoon attaches itself to the ovum. The head and middle piece pass into the cytoplasm of the egg, leaving the tail behind. The ovum now contains two nuclei and two centrosomes, belonging to the spermatozoon and ovum respectively. The two centrosomes approach one another, and between them is formed a spindle upon which the chromosomes of the egg-nucleus and those of the sperm-nucleus are distributed. In some animals the paternal and maternal chromosomes are distinct, and in such cases it can be seen that each set divides, and the four groups go to opposite poles of the spindle—two to each. The whole cell, or *zygote*, now divides, so that each daughter-cell receives an equal number of both paternal and maternal chromosomes. These being reduced by the changes described on pp. 168-9, now by addition amount to the full number characteristic of the species.

INDEX.

[*References to development are printed in thick numerals.*]

N

O

P

Printed by R. & R. CLARK, LIMITED, *Edinburgh*.